John Parker

Elementary Thermodynamics

John Parker

Elementary Thermodynamics

ELEMENTARY THERMODYNAMICS

BY

J. PARKER, M.A.,
FELLOW OF ST JOHN'S COLLEGE, CAMBRIDGE.

CAMBRIDGE:
AT THE UNIVERSITY PRESS.
1891

[All Rights reserved.]

Cambridge:
PRINTED BY C. J. CLAY, M.A. AND SONS,
AT THE UNIVERSITY PRESS.

PRINTED IN GREAT BRITAIN

NOTE.

BEGINNERS may omit the following parts:—
    Chapter I., Arts. 20—25.
    Chapter III., Arts. 65, 72—90.

The word 'Elementary' is used in the title of this book because it does not enter into the details of electricity and magnetism.

CONTENTS.

CHAPTER I.

	PAGES
THE CONSERVATION OF ENERGY	1–80

CHAPTER II.

ON PERFECT GASES	81–103

CHAPTER III.

CARNOT'S PRINCIPLE	104–236

CHAPTER IV.

APPLICATIONS OF CARNOT'S PRINCIPLE	237–324

CHAPTER V.

THE THERMODYNAMIC POTENTIAL	325–343

CHAPTER VI.

	PAGES
APPLICATIONS OF THE THERMODYNAMIC POTENTIAL	344–378
NOTE A	379–381
NOTE B	382–391
APPENDIX . .	393–408

ELEMENTARY THERMODYNAMICS.

CHAPTER I.

THE CONSERVATION OF ENERGY.

1. Units of Measurement.—As it is now becoming universal to express all dynamical quantities in terms of a very convenient set of units based on the Metric, or French, system of weights and measures, we shall give an account of the method, and also make a comparison with the less simple, but more familiar, English system.

In the metric system, which was established in France by law in 1795, and is now widely adopted for commercial purposes, the standards of length and mass are the Metre and the Kilogramme, respectively. The Metre is the distance between the ends of a rod of platinum made by Borda, when the temperature is that of melting ice: the Kilogramme is the mass of a piece of platinum, also made by Borda.

The subsidiary measures of the metric system are formed as follows:—

1 kilometre = 1000 metres.	1 kilogramme = 1000 grammes.
1 hectometre = 100 metres.	1 hectogramme = 100 grammes.
1 decametre = 10 metres.	1 decagramme = 10 grammes.
1 decimetre = $\frac{1}{10}$ metre.	1 decigramme = $\frac{1}{10}$ gramme.
1 centimetre = $\frac{1}{100}$ metre.	1 centigramme = $\frac{1}{100}$ gramme.
1 millimetre = $\frac{1}{1000}$ metre.	1 milligramme = $\frac{1}{1000}$ gramme.

P.

The Litre (used for liquids) is the same as the cubic decimetre.

The standards of the metric system were originally chosen so that the metre should be the ten-millionth part of the distance from the pole to the equator, and the kilogramme the mass of a litre of distilled water, at 4° C., the temperature of maximum density, as nearly as could be then determined.

In the centimetre-gramme-second, or C.G.S., system of absolute units, now so generally used in dynamics, the fundamental units of length, mass, and time, are chosen to be the centimetre, gramme, and mean solar second, respectively. The C.G.S. absolute unit of force, called a Dyne, is then defined to be that force which, acting for one second on the mass of a gramme, will generate a velocity of one centimetre per second. The C.G.S. absolute unit of pressure is a pressure of one dyne per square centimetre.

In conjunction with these absolute units, we frequently employ arbitrary units, as the gramme-weight for force, and the millimetre of mercury and the atmo for pressures.

As the weight of a given mass is not quite the same in all parts of the world, whenever we speak of a gramme-weight as a measure of force, we shall mean a force equal to the weight of a gramme at Paris. The acceleration of any body falling freely at Paris under its own weight being 980·868 centimetres per second, we see that the weight of a gramme at Paris is 980·868 dynes. To determine the weight of a gramme at any other place, we may use Clairaut's formula

$$g = G\left(1 - \cdot 0025659 \cos 2\lambda\right)\left(1 - 1\cdot 32 \frac{z}{r}\right),$$

where λ is the latitude of the place, z its height above the mean level of the sea, r the radius of the earth, g the number of dynes in a gramme-weight at the place, and G the value of g at the mean level of the sea in latitude $45°$.

The Atmo is defined to be the pressure produced (at Paris) by a column of mercury 760 millimetres high, when the temperature is $0°$ C., and the top of the mercury subjected to no force but the pressure of its own vapour. It is found that an atmo is equal to the weight (at Paris) of 1033·279 grammes per square centimetre, or to a pressure of 1013510 dynes per square centimetre.

The Atmo, the name of which is due to Prof. J. Thomson, is about equal to the average pressure of the atmosphere in ordinary places and is chiefly used for measuring high pressures: the gramme-weight is very convenient in estimating small forces.

In the English, or practical, system of dynamical units, the fundamental units of length, mass, and time, are the foot, the pound avoirdupois, and the mean solar second, respectively. The absolute unit of force, which Prof. J. Thomson calls a Poundal, is that force which, if it acted for a second on the mass of a pound, would generate in it a velocity of one foot per second. The accelerating effect of gravity being 32·1889 feet per second at London, it follows that the weight of a pound at London is 32·1889 poundals. Hence, for rough purposes, we may consider a poundal equal to the weight of half an ounce in any part of the world.

We shall always work in the C.G.S. system of units, but it will be easy to express our results in the English practical method by means of the following data:

4 ELEMENTARY THERMODYNAMICS.

1 metre = 39·370432 inches.   1 inch = 2·54 centimetres,*
1 centimetre = ·393704 inch.   1 foot = 30·48 centimetres,*
1 sq. metre = 1550·03 sq. in.   1 sq. in. = 6·45 sq. centimetres,*

* Nearly.

1 litre = 61·025 cubic inches = 1·7617 pint.
1 pint = ·5676 litre.
1 kilogramme = 2·2046 lbs. avoir. = 15432 grains.

Hence we have, roughly:—

Weight of 1 lb. avoir. = 444900 dynes.
1 lb. per sq. inch = 70·3 grammes per sq. centimetre = 69000 dynes per sq. centimetre.
1 gramme per sq. centimetre = ·0142 lb. per sq. inch.
1 atmo = 14·7 lbs. per sq. inch.

2. The Conservation of Energy.—We are now prepared to explain the principle of the Conservation of Energy. This is a deduction, by means of experiment and experience, from a more obvious result known as the principle of Work, which will now be obtained as a direct theoretical consequence of Newton's three Laws of Motion.

In order to apply these laws correctly, it is necessary to conceive the bodies which we are considering to be made up of a large number of very small pieces, or *particles*, each of which is so small in all its dimensions that for all dynamical purposes it may be treated as a mathematical point; the relative motions of its parts with respect to one another being negligible in comparison with its motion of translation as a whole.

Before we are able to consider real bodies of finite size, we must take the case of a single particle.

3. When a particle, which is moving about in any manner under the action of a force of p dynes, receives a small displacement of *ds* centimetres, in a direction which is not at right angles to the force, the force is said to do Work on the particle, or the particle is said to do work against the force, and the amount of work done on the particle is defined to be the product of the displacement into the resolved part of the force along the displacement. If ϵ be the angle between the positive directions of p and *ds*, the work done by the force will therefore be $p \cos \epsilon \cdot ds$, and may obviously be either positive or negative. It is also clear that the work done by the force is equal to the product of the force and the projection of the displacement along the force.

The absolute unit of work in the C.G.S. system of units is called an Erg, and is the work done by a force of one dyne when it moves its point of application one centimetre in its positive direction. The absolute unit of work in the English system is the Foot-poundal, which is the work done by a force of one poundal when its point of application moves one foot in the positive direction of the force.

Work is also reckoned in arbitrary units, as the gramme-centimetre and the foot-pound. These different methods may easily be compared by means of the following relations:

    1 gramme-centimetre (at Paris) = 980·868 ergs.
    1 foot-pound = 1356×10^4 ergs, roughly.

If several forces act simultaneously on the particle, the work which they do is defined to be the work done by their resultant. And since the resolved part of the resultant along the displacement is equal to the sum of the resolved parts of the component forces in the same direction, it is evident the work done by the resultant is equal to the sum of the quantities of work done by the forces separately. Thus if R be the sum of the resolved parts of the forces along the displacement ds, and dW the work done on the particle, we have $dW = Rds$.

To put this into analytical language, let (X, Y, Z) be the sums of the components of the forces which act on the particle, parallel to three fixed rectangular axes (Ox, Oy, Oz), with which the displacement ds makes angles (α, β, γ), respectively. Then, by the principles of statics,

$$R = X \cos \alpha + Y \cos \beta + Z \cos \gamma.$$

But if (dx, dy, dz) be the projections of the displacement ds on the axes, we have, since the axes are rectangular,

$$dx = ds \cos \alpha, \qquad dy = ds \cos \beta, \qquad dz = ds \cos \gamma.$$

Hence $\qquad dW \equiv Rds = Xdx + Ydy + Zdz \ldots\ldots\ldots(1).$

If the particle move from an initial position A, whose coordinates are (x_0, y_0, z_0), to a final position B, whose coordinates are (x_1, y_1, z_1), the work done on it by the forces may be written

$$\int_0^1 dW \quad \text{or} \quad \int_0^1 (Xdx + Ydy + Zdz),$$

where the integral sign simply denotes a summation, without implying that (x, y, z) can be taken as independent variables.

4. Again, however complicated may be the connections of the given particle with other particles, we may always, for dynamical purposes, treat it as a perfectly free particle whose motions depend only on the forces which act upon it. We may therefore take the time as the

THE CONSERVATION OF ENERGY.

independent variable. Hence, if m be the mass of the particle in grammes, and (u, v, w) its velocities parallel to the axes in centimetres per second, we have, by the second law of motion

$$m\frac{du}{dt} = X, \quad m\frac{dv}{dt} = Y, \quad m\frac{dw}{dt} = Z.$$

Thus, since
$$dx = udt, \quad dy = vdt, \quad dz = wdt,$$
we obtain
$$m\left(u\frac{du}{dt} + v\frac{dv}{dt} + w\frac{dw}{dt}\right)dt = Xdx + Ydy + Zdz,$$

and therefore when the particle moves from (x_0, y_0, z_0) to (x_1, y_1, z_1), we shall have

$$\tfrac{1}{2}m(u_1^2 + v_1^2 + w_1^2) - \tfrac{1}{2}m(u_0^2 + v_0^2 + w_0^2)$$
$$= \int_0^1 (Xdx + Ydy + Zdz).$$

But, since the axes are rectangular,
$$Xdx + Ydy + Zdz = dW,$$
and if V be the total or resultant velocity of the particle at any instant, we have $V^2 = u^2 + v^2 + w^2$, so that

$$\tfrac{1}{2}mV_1^2 - \tfrac{1}{2}mV_0^2 = \int_0^1 dW \dots\dots\dots\dots(2).$$

Now the Kinetic Energy of a particle is defined to be half the product of its mass and the square of its velocity. The last result therefore shows that as the particle moves about, the algebraic increase of its kinetic energy is equal to the work done upon it.

It is clear that kinetic energy is measured in the same way as work, that is, in ergs (or foot-poundals).

5. The forces which act on the particle are due, partly at least, to the influence of other particles, and as these may themselves be moving about, it is clear that the components (X, Y, Z) will not necessarily depend on the position of the given particle alone. Hence, when the particle moves from A to B, the work done upon it and therefore also the change in its kinetic energy, will not generally depend solely on the positions of A and B, and on the path along which the motion takes place. But if (X, Y, Z) depend only on the position of the given particle, they will be functions of its coordinates (x, y, z), and the work done by the forces will depend only on the two positions A, B, and on the path by which the particle travels from one to the other. In this case, the integral

$$\int_0^1 dW, \text{ or } \int_0^1 (X dx + Y dy + Z dz),$$

falls under one or other of two heads between which there is a very important distinction, illustrating a point of frequent occurrence in thermodynamics.

First, if (X, Y, Z) satisfy the three relations

$$\frac{dX}{dy} = \frac{dY}{dx}, \quad \frac{dY}{dz} = \frac{dZ}{dy}, \quad \frac{dZ}{dx} = \frac{dX}{dz} \quad \ldots\ldots\ldots\ldots(3),$$

then it can be shown that the expression

$$X dx + Y dy + Z dz, \text{ or } dW,$$

is the complete differential of some function of (x, y, z), so that we may write

$$W = F(x, y, z) + C,$$

where C is an arbitrary constant.

Hence if the particle move from the position A to the position B, the work done by the forces takes the form

$$F(x_1, y_1, z_1) - F(x_0, y_0, z_0),$$

and therefore when the last position is the same as the first, the work done on the particle will be

$$F(x_0, y_0, z_0) - F(x_0, y_0, z_0).$$

Now if the function F contain such terms as $\tan^{-1} \dfrac{z^2}{xy}$, which may have different values for the same values of (x, y, z), the last expression will not necessarily be zero. But if we suppose $F(x, y, z)$ to be such that it has a single value for each point in space, in other words, that it is a single-valued function, then the results just obtained show that when the particle moves from one position to another, the work done by the forces depends only on the initial and final positions, and is therefore the same by whatever path the particle may have travelled from one to the other: also that when the last position coincides with the first, the work done is zero.

Secondly, when the three conditions in equation (3) are not all satisfied, the expression for dW will not be a complete differential and the integration will be impossible until something further is specified about the path. This will be made clear by an example.

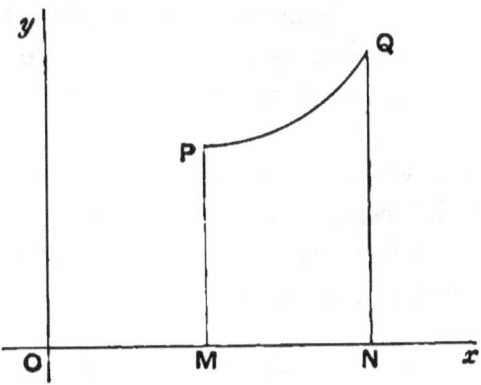

Let us suppose that $X = \mu y$, where μ is a constant, and that both Y and Z are constantly zero. Then we have

$$dW = \mu y\, dx.$$

If then the curve PQ be the projection on the plane of (xy) of the path of the particle, and PM, QN perpendiculars from P and Q on Ox, the work done on the particle by the forces will be μ times the area of the figure $PQNM$, which will depend on the form of the curve PQ as well as on the positions of the points P, Q.

In general, if a particle pass from any position A to any other position B, under the influence of forces which depend only on the position of the particle but do not make the expression $Xdx + Ydy + Zdz$, or dW, a complete differential, the work done by the forces will depend not only on the two positions A, B, but also on the path by which the particle travels from A to B. Hence if the particle return from B to A under the influence of the same forces but by a different path, the work done in the return path will not necessarily be equal and opposite to that done in the first case. The total work done will therefore generally be different from zero. Hence if the particle pass through the same point several times, its velocity will generally be different every time.

Again, even when no function W exists of which dW is the exact differential, it is still found convenient to employ the ordinary notation of partial differential coefficients. Thus the equation

$$dW = Xdx + Ydy + Zdz$$

gives, when dy and dz are both zero,
$$dW = Xdx,$$
which we shall write in the form $\left(\dfrac{dW}{dx}\right) = X$. Similarly we may write $\left(\dfrac{dW}{dy}\right) = Y$ and $\left(\dfrac{dW}{dz}\right) = Z$; but we must remember that $\dfrac{d}{dx}\left(\dfrac{dW}{dy}\right)$ is not equal to $\dfrac{d}{dy}\left(\dfrac{dW}{dx}\right)$, &c., &c. In fact, we have just seen that these are the very conditions that make dW an exact differential.

6. A system of real bodies is to be considered as an assemblage of a vast number of particles, and its kinetic energy and the work upon it in any given time are defined to be the sums of the corresponding quantities for the individual particles. Hence if T be the kinetic energy of the system at any instant, $T + dT$ the kinetic energy at a consecutive instant, and dW the work done on the system in the interval, we shall have
$$dT = dW.$$

This equation has been simplified by the decisive discovery of the existence of the Ether. This is a frictionless substance or medium pervading all space which is so subtle that it cannot be removed from a vessel in the process of forming a 'vacuum.' Not being directly apparent to our senses, it is not considered to form part of our material system. Of the various functions which it is known to perform in nature, one of the most obvious is the transmission of light and heat across the vacant space which separates the sun from the earth. This was discovered by Roemer in 1675 to be a gradual and not an instantaneous process, the velocity of light in a vacuum being about 299860[1] kilometres, or 186326 miles, per second. The fact that the presence of the ether does not sensibly impede the motions of the planets is illustrated by a well-known proposition in hydrodynamics, that a body of invariable form experiences no resistance in moving through a frictionless fluid with uniform velocity in a straight line.

According to modern ideas, the forces which act on any particle consist of pushes and pulls due to its direct contact with other particles or with the ether. Forces may therefore be distinguished as contact-forces and ether-forces, respectively, and if dW_c be the work done on the whole system by the contact-forces and dW_e the work done by the ether-forces, we shall have

$$dT \equiv dW = dW_c + dW_e.$$

But when any two particles are in contact, their mutual contact-forces are equal and opposite, and as soon as the particles separate, the forces both vanish. Thus so long as the forces exist, their points of application are coincident, and therefore the quantities of work which they do in any the same time are equal and opposite, or the sum of the quantities of work is zero. Hence, for the whole system, the work done by the internal contact-forces in any given time is zero, and therefore if dw_c be the work done by the external contact-forces, we have

$$dw_c = dW_c,$$

so that the principle of work takes the final form

$$dT = dw_c + dW_e \dotfill (4).$$

---

[1] This is Newcomb's determination, given in Everett's 'Units and Physical Constants.'

7. The value of the principle of energy arises from the fact that all the ether-forces in nature are such as to give to dW_e a very simple form. The methods by which this important result has been established are of two kinds. First, there are a number of experiments which make it appear extremely probable, and secondly, whenever the principle of energy has been made use of in the study of physical phenomena, it has been found to lead to consequences which are in all cases in exact agreement with observation; indeed, the theoretical calculations have frequently been performed first and the experimental verifications obtained afterwards.

If all the different varieties of ether-forces were sufficiently understood by us, we should probably be able to give a general theoretical proof of the principle of Energy. Unfortunately, we know very little about forces, especially about those which exist in the interior of solid and fluid bodies; still there are a few forces which can be dealt with by theory alone. We shall therefore adopt the following method of treatment :—First of all, the principle will be proved for a few ideal systems in which the only ether-forces which are supposed to exist, or to do work, are identical with certain very simple forces which are of constant occurrence in real systems. After this, some of the peculiarities of other ether-forces will be noticed; by means of which a clear general conception of the principle will be gained. We are then immediately led to a very simple theoretical explanation of the nature of heat. Lastly, the experimental evidence will be considered, and it will be seen to be a mere repetition of the theoretical arguments as to the nature of heat.

8. In the first place, let us suppose that there are no ether-forces which do work. Then if dw be the work done by the external contact-forces, we have

$$dT = dw,$$

and therefore when the external work is zero, the kinetic energy is constant, so that the only effect of the internal forces is a transference of kinetic energy from one part of the system to another.

9. Next, suppose the only ether-forces are due to gravitation, so that, according to Newton's discovery, two particles whose masses are m and m' grammes and whose distance is r centimetres, attract one another with equal and opposite forces of $\lambda \dfrac{mm'}{r^2}$ dynes, where λ is a constant number which is easily determined from astronomical and other considerations to be about $\dfrac{6\cdot 48}{10^8}$.

Gravitation belongs to a class of forces which are known as actions at a distance, because they exist between particles when they are not in contact with one another, and even when they are separated by an ordinary vacuum, as we may see in the case of the attractions between the heavenly bodies. It was formerly supposed that actions at a distance were exerted across nothing, but it is now believed that they are due to continuous contact-forces in the ether.

To find the work done by the internal gravitational forces, we take them in pairs. Suppose, then, that P, Q are the positions of two particles of masses (m, m') when their distance is r, and let P', Q' be neighbouring positions such that $P'Q' = r + dr$. Drop the perpendiculars PM

THE CONSERVATION OF ENERGY.   15

and QN from P and Q on $P'Q'$. Then if dw_1 be the work

done by the mutual attraction on the particle at P and dw_2 the work done on the particle at Q, we have

$$\left. \begin{aligned} dw_1 &= -\lambda \frac{mm'}{r^2} \cdot P'M \\ dw_2 &= -\lambda \frac{mm'}{r^2} \cdot Q'N \end{aligned} \right\},$$

and therefore

$$dw_1 + dw_2 = -\lambda \frac{mm'}{r^2}(P'M + Q'N) = -\lambda \frac{mm'}{r^2}(P'Q' - MN).$$

But if ϵ be the small angle between the directions of PQ and $P'Q'$,

$$MN = r \cos \epsilon = r,$$

small quantities of the second order being rejected.

Thus $$dw_1 + dw_2 = -\lambda \frac{mm'}{r^2} dr,$$

and the work done on the system by all the internal gravitational forces may be written

$$-\lambda \Sigma \frac{mm'}{r^2} dr,$$

which is obviously a perfect differential. Hence when the system passes from a state O to another state P, the work done on it by the internal forces will be the same for all paths and may be written W_{op}, since it depends only on the two states O, P. Equation (4) therefore gives

$$T_P - T_O = \int_O^P dw_c + \int_O^P dw_g + W_{OP},$$

where $\int_O^P dw_c$ and $\int_O^P dw_g$ are the quantities of work done by the external contact and gravitational forces, and the suffixes refer to the respective states. For shortness, this result may be written

$$T_P - T_O - W_{OP} = \int_O^P dw,$$

and $\int_O^P dw$ may be called the external work done on the system. If the state O be *fixed*, the term $-W_{OP}$ will only vary with the state P, and may therefore be written V_P:

hence $$T_P + V_P - T_O = \int_O^P dw \quad \ldots\ldots\ldots\ldots(5).$$

It thus appears that if no external work be done on the system, $T + V$ remains constant, so that T can only increase at the expense of V, and vice versâ. For this reason V is regarded as a second kind of energy, and has received the appropriate name of Potential Energy from Rankine, to indicate that it is convertible into kinetic energy. Its value in any state P is equal to the work which the internal forces do on the system as it returns in any manner from the state P to any other state in which the relative positions of the particles are the same as in the standard fixed state O, so that in the standard state the potential energy is zero.

It was formerly thought that kinetic energy was actually created when it increased at the expense of the potential energy in the case of a system under the action of no external forces, but it is now supposed to be merely transferred from the ether.

The sum of the kinetic and potential energies is called the 'Internal Energy,' or, more shortly, the 'Energy' of the system, and is usually written U[1], so that equation (5) becomes

$$U_P - U_O = \int_O^P dw \quad \ldots\ldots\ldots\ldots\ldots(6).$$

If we take a different state O' for the standard fixed state and denote the new value of U_P by U'_P, we have

$$\left. \begin{array}{l} U'_P \equiv T_P + V'_P = T_P - W_{O'P} \\ U_P = T_P - W_{OP} \end{array} \right\},$$

whence $\qquad U'_P - U_P = W_{OP} - W_{O'P}.$

Now since the work done by the internal forces is the same for all paths, we may suppose the path $O'P$ to pass through O. Consequently

$$W_{O'P} = W_{O'O} + W_{OP},$$

and therefore $\qquad U'_P - U_P = - W_{O'O},$

which is independent of the state P.

Hence, since it is clear that the energy cannot have two different values corresponding to the same state of the system, it follows that U is a single-valued function of the independent variables which define the state of the system, together with an arbitrary additive constant depending only on the choice of the standard state.

---

[1] Sometimes, in English books, the letter E is used to denote internal energy. The notation adopted in the text appears to be preferable, because E is required in Electricity to denote 'electromotive force.'

Thus if $(x, y, z, \ldots\ldots)$ be the independent variables, we have
$$U = f(x, y, z, \ldots\ldots) + C,$$
where C is an arbitrary constant.

If we take the difference between the values of U in two different states of the system, the arbitrary constant will not appear in the result, which is therefore perfectly definite.

When the system consists of a number of separate bodies, its energy depends not only on the state of each of its parts but also on their relative positions with respect to one another. The energy which depends on the relative positions of the bodies is called their 'mutual' potential energy, or, shortly, their 'mutual energy,' and is evidently the excess of the energy of the whole system over the sum of the internal energies of the bodies of which it is composed.

10. Again, let us suppose our system to be influenced by radiation as well as by contact-forces and gravitation. According to modern theories, radiation consists of spherical waves of motion in the ether which are excited by the irregular vibrations of the smallest parts, or 'atoms[1],' of matter, somewhat after the manner of the circular waves which may be produced by dropping a stone into still water. The reasoning on which this conclusion is based involves optical principles which cannot be discussed in this book, but it will become evident later on that all bodies are in a state of vibration. Thus a wave of radia-

---

[1] An 'atom' is a chemical reality, a particle a pure mathematical conception for the purpose of calculation. An atom generally contains an infinite number of particles.

tion possesses energy, but its energy differs from the potential energy of gravitation in not being bound to material bodies and carried about with them.

When a wave of radiation falls on a material system, it will affect both the potential and kinetic energy of the system; but the effect on the potential energy is generally exceedingly minute. Hence, to bring the subject of radiation within the scope of this book, we may *assume* that the effect of radiation on a system is produced *entirely* by the imperceptible impacts of the waves on the material particles of the system[1]. If dw_r be the work done on the system by these forces, which may, for shortness, be called radiation-forces[2], equation (4) becomes

$$T_p - T_0 = \int_0^P (dw_c + dw_g + dw_r) + W_{op}.$$

Putting dw for $dw_c + dw_g + dw_r$ and supposing the state O to be fixed, this result may be written in any of the forms

$$\left. \begin{array}{l} T + V_p - T_0 = \int_0^P dw \\ U_p - U_0 = \int_0^P dw \\ dU = dw \end{array} \right\} \quad \ldots\ldots\ldots\ldots(7).$$

The energy which is taken from the system by gravitation or radiation passes into the ether, but the energy

---

[1] That is, when a body is absorbing radiation, the energy so gained is equal to the work done on the material particles of the body by the incident waves; when a body is emitting radiation, the energy so lost is equal to the work done on the ether by the vibrating particles in starting the waves.

[2] As will be seen later, the particles of a body vibrate in such different directions, that the resultant of the radiation-forces which act on them will generally be quite imperceptible.

which is lost owing to contact-forces passes directly into some other system. Hence, when the external forces which act on a system are all contact-forces, the external work is frequently referred to as the work done on the system by external *bodies*.

Radiation-forces are, of course, far too small to be detected by instrumental means; nevertheless the work $\int_0^P$ can be found. For if the change from the state O to the state P be effected by means of a measured quantity of work, W, done by contact-forces and gravitation alone, we have

$$\int_0^P dw = W.$$

The simple fact we have just obtained for our ideal systems, that the energy U increases by the amount of external work done on the system, or that dw is a perfect differential, is the great principle of the Conservation of Energy. It appears to be true for all systems found in nature, and in consequence all natural forces are said to be Conservative.

11. The principal ether-forces in nature which do work, in addition to gravitation and radiation-forces, are those which give rise to chemical, physical, electric, and magnetic actions.

When a chemical or physical process takes place in a system, there is a large amount of work done by the ether-forces in the interior of the system; but it is found that no consequent effect is produced where the process does not actually take place, except by contact-forces and radiation. Hence, if there are no electric or magnetic actions, our system is necessarily so chosen that the only external forces are contact-forces, gravitation, and radiation-forces. It is then found, by experiment and experience, that the system satisfies a relation of the form

$$U_P - U_0 = \int_0^P dw,$$

or
$$dU = dw,$$

where U is a single-valued function of the independent variables which define the state of the system together with an arbitrary additive constant depending only on the choice of the standard fixed state O. The potential energy, it must be remembered, is not the same as if gravitation were the only internal force.

Electric and magnetic forces are the only ether-forces besides gravitation about which much is known, for the simple reason that they are the only other ether-forces which have been measured. One of the chief peculiarities in which they differ from gravitation is that the properties of exerting actions at a distance may be transferred from one particle to another. In consequence of this, it is found that a system may acquire such an electric condition by contact with external bodies that it is impossible to bring it into the standard state until the new electric properties are given back. We are then to regard the system as a new system and to choose a state into which the system can be brought as a new standard state from which its potential energy may be reckoned. The difficulty may be obviated by extending our system so as to include the bodies from which these electric properties have been obtained. If our system be so chosen, and if there be no external electric or magnetic forces, then it is found, however complicated may be the chemical or electric actions in its interior, that it satisfies the relation

$$U_P - U_0 = \int_0^P dw,$$

or
$$dU = dw.$$

In applying the principle of energy to an electrified or magnetized system, there is a special consideration to be attended to. Thus let P, Q be two different states of the same system which can be compared with the same standard state or origin. Suppose also that the kinetic energies of the particles are the same in both cases, but that the potential energy is greater in the state Q than in the state P. Then it is found that when there are other electrified or magnetized bodies in the neighbourhood, the system may be brought from the state P to the state Q without doing external work. But though the external forces do no work on the material part of the system, they do work on that part of the ether which, according to one of Fresnel's great discoveries, is inseparably connected with the system. In fact, it appears from works on Electricity, that, under these circumstances, the bound-ether may be regarded as a spring which can be bent independently of the material part of the system. If then dU be the increase of the energy in any change of state and dw the external work done on the material particles of the system, we must write

$$dU = dw + de,$$

where de may be called the 'electric work' done on the system.

Again, the system may be acquiring electric properties in one part of its mass and simultaneously losing in another in such a way that the potential energy may always be reckoned from the same standard state or origin. When, as in the case of electric currents, the properties which the system gains bring with them more potential energy than those which it loses take away, the increase of energy will be different from the external work done on the material part of the system, and we must again write

$$dU = dw + de.$$

From this it is evident that electric and magnetic actions occupy a place between gravitation, on the one hand, and radiant, or free etherial, energy, on the other.

As we do not intend to enter into the details of electric and magnetic forces in this work, we shall always suppose our system so chosen that the only external forces are contact-forces, gravitation, and radiation-forces, except when it is expressly stated otherwise. The principle of energy may then be stated in the simple form

$$U_P - U_O = \int_O^P dw,$$

or
$$dU = dw,$$

so that dw is a complete differential.

It is obvious that the only effect of the internal ether-forces is a transformation of kinetic into an equal amount of potential, energy, and vice versâ.

Since the principle of energy does not require us to consider the internal forces, we shall frequently drop the adjective in referring to the external forces or work. The energy of a system may then be described as its capacity for doing work, and positive work may be regarded as the passage of energy into the system.

We now proceed to distinguish kinetic energy and work into their visible and invisible parts, so as to explain the phenomena of Heat.

12. A very little observation enables us to perceive a general tendency in nature for all solid bodies, or for liquids and gases contained in vessels, to assume a sensibly invariable form and internal condition, except when they are prevented by external causes (including as such the radiation of energy into external space). Thus if we strike a bell or other body, vibrations are produced which are frequently visible to the eye, but always disappear from sight more or less rapidly. Indeed, after a sufficient time has elapsed, the most powerful microscope fails to detect any vibrations in the body. Again, if two moving bodies collide, they may be eventually brought nearly to a state of apparent rest by the collision, so that there will seem to be a considerable loss of kinetic energy without a corresponding increase of potential energy. And since many cases occur in which we naturally suppose very little energy to be lost from the two colliding bodies before the vibrations subside, we are led to conjecture that kinetic energy may exist in an invisible as well as in a visible and palpable form. It will be seen hereafter that there are experiments which raise this suspicion to a certainty. We therefore define the Mechanical Kinetic Energy of a system of bodies to be the kinetic energy that the system would have if its motions were the same as they appear to be, or, more exactly:—

If we divide a material system into a large number of parts, and then multiply the mass of each by the square of the velocity of its centre of mass and take the sum, the result obtained will approach a limit as the number of parts is continually increased; but after this limit is practically reached, it may begin to diverge from it

THE CONSERVATION OF ENERGY. 25

and finally arrive at another limit, which is the true kinetic energy of the system. When there are two or more limits, the first of the improper limits is defined to be the Mechanical Kinetic Energy of the system, and the excess of the true limit over this, the Non-mechanical Kinetic Energy. Since it will appear that kinetic energy is always present in the invisible form, it is clear that if there is only one limit in the above process, the mechanical kinetic energy is zero and the system in a state of mechanical rest.

It can easily be shown that the mechanical kinetic energy is always less than the total kinetic energy, and, consequently, the non-mechanical kinetic energy always positive. Thus let M be the mass of one of the parts into which the system has just been divided, (u, v, w) the velocities of its centre of mass parallel to three rectangular axes, $(m_1 m_2 m_3 ...)$ the masses of its ultimate particles and $(u_1 v_1 w_1)$, $(u_2 v_2 w_2)$,... their velocities parallel to the same axes. Then we have, by the conservation of linear momentum,

$$m_1 u_1 + m_2 u_2 + m_3 u_3 + \ldots = Mu,$$

whence

$$m_1^2 u_1^2 + m_2^2 u_2^2 + m_3^2 u_3^2 + \ldots + 2 m_1 m_2 u_1 u_2 + \ldots = M^2 u^2.$$

Hence $M^2 u^2$ cannot be greater than

$$m_1^2 u_1^2 + m_2^2 u_2^2 + m_3^2 u_3^2 + \ldots + m_1 m_2 (u_1^2 + u_2^2) + \ldots,$$

or than

$$(m_1 + m_2 + m_3 + \ldots)(m_1 u_1^2 + m_2 u_2^2 + m_3 u_3^2 + \ldots).$$

Thus $\frac{1}{2} M u^2$ cannot be greater than $\frac{1}{2} \Sigma m_1 u_1^2$; and similarly for the velocities parallel to the other axes; which proves the proposition.

The non-mechanical kinetic energy of a system is

supposed to be that part of its energy on which its thermal properties depend.

It should be noticed that in the case of elastic bodies, like iron or compressed gases, the capacity of acquiring mechanical kinetic energy without the assistance of external forces—a property which will hereafter be defined as mechanical potential energy—is not necessarily due, like true potential energy, to the existence of internal ether-forces.

13. In consequence of non-mechanical kinetic energy, work may be done by contact-forces in an invisible as well as in a visible manner. Suppose, for example, that any two bodies A, B are in contact and that the surface of contact is apparently at rest. Then no visible work is done by the contact-forces; but if the surface particles of A and B remain in contact through a much greater distance when the direction of their vibratory motion is from A towards B than when it is in the contrary direction, the total work of the contact-forces done by A upon B may be considerable. We therefore define the Mechanical Work done upon any system of bodies by the external contact-forces to be the work that they would actually do if the motions of the surfaces of contact were the same as they appear to be, or, more exactly:—

If, in order to find the work done on any system in any small change of state by the contact-forces due to external bodies, we divide the surfaces of contact into a large number of parts, and then multiply the displacement of the centre of each of these small areas by the resolved part in the direction of the displacement of the force which acts upon it and take the sum, the result obtained will approach a limit as the number of parts is

THE CONSERVATION OF ENERGY. 27

continually increased; but after this limit is practically reached, it may begin to diverge from it and finally arrive at another limit, which is the true work done by the contact-forces on the system. When there are two or more limits, the first of the improper limits is defined to be the Mechanical Work, and the excess of the true limit over this, the Non-mechanical Work. If there is only one limit, either the mechanical or the non-mechanical, work is zero.

Since every force which acts upon the system produces a change in some, at least, of the independent variables (x, y, z,...) which define the state of the system, it is evident that both the mechanical and the non-mechanical work done on the system in any small change of state are functions of the independent variables and of their differentials.

It appears from experiment that the non-mechanical work done by contact-forces is what we understand when we speak of the conduction of heat.

In the case of gravitation, the force between any two particles depends only on their distance and therefore cannot change abruptly, like a contact-force. From this it can be shown that the whole of the work done on any system by the external gravitational forces is mechanical work, and that, in calculating it, we need take no account of the non-mechanical motions. This result, joined to the fact that an enormous amount of 'heat' is transmitted from the sun to the earth, affords a strong argument for the existence of the ether.

Thus, if dU be the increase of the energy of any system in any small change of state, and dw the total work done upon it, consisting of a quantity dW of mechanical work and a quantity dQ of non-mechanical work, due either to the conduction of heat or to radiation, we have

$$dU = dw = dW + dQ \quad\ldots\ldots\ldots\ldots\ldots(8),$$

from which we draw the important conclusion that dW and dQ will either both simultaneously be, or both not be, complete differentials of functions of the independent variables.

14. If the surface of the smoothest body be examined by a powerful microscope, it is found to be so irregular that it may be said to be covered with a great number of small projecting teeth. Besides the irregularities revealed by the microscope, there are probably a vast number too small to be detected. Hence, if any two bodies, A, B be pressed together, their surfaces of contact will sink into one another, and if we attempt to move one body over the other, we shall experience a resistance in addition to the external forces. This resistance is at once recognised as Friction, and by supposing the common surface of A, B to be plane, and taking account of the vast number of teeth found even on the smallest area, we may easily obtain some of its chief 'Laws,' thus:—

I. Friction acts on each body in a direction opposite to that in which its relative motion takes place, or merely tends to take place.

II. No more friction can ever be called into play than is just sufficient to prevent relative motion; but since the amount of friction cannot be infinite, it is clear that the frictional resistance will be unable to prevent relative motion when the force which tends to produce it is large enough. Hence, as the force which tends to cause relative motion continually increases from zero, friction will continually increase with it, at least until the common surfaces begin to slip. This is usually expressed by saying that if friction can prevent relative motion, it will. The amount of friction called into play between any two given bodies by a given normal pressure when slipping is about to take place, is called the 'limiting' friction for that normal pressure.

III. Let C, D be two other bodies in contact whose natures and conditions are the same as those of A, B, respectively, and suppose the total normal pressure between C, D to be n times as great as between A, B. Then if the common area of C, D be n times the common area of A, B, and if the normal pressures be evenly distributed in both cases, it is clear that, for these normal pressures, the limiting friction between C, D will be n times as great as the limiting friction between A, B. From experiment it further appears, at least very approximately, that when the total normal pressures are as n to 1, the limiting amounts of friction will be in this ratio whatever be the proportion of the common areas. Hence, if R be the total normal pressure between any two given bodies whose common surface is plane, the limiting friction between them will be μR, where μ is a constant number (called the coefficient of friction) depending on the natures and states of the two bodies, but independent both of the total normal pressure and of the area of the surface of contact.

15. On account of friction, an expenditure of mechanical work is always necessary in order to cause two bodies to slide over one another. If, as often happens, the two bodies possess no appreciable potential energy, or, at least, only a constant amount, and if the mechanical kinetic energy be zero both before and after slipping, the only effect on the two bodies of the mechanical work done on them will be an increase in their non-mechanical kinetic energy. Now it is evident that exactly the same effect might have been produced by doing an equal amount of non-mechanical work. Thus the same change of state may be brought about in the two given bodies by means of mechanical work alone, or by means of non-mechanical work alone, or partly by means of mechanical and partly by means of non-mechanical, work, in any given ratio. Hence, on account of friction, when any system experiences a change of state, the quantities of mechanical and of non-mechanical work done upon it will generally depend on the way in which the change of state takes place as well as on the initial and final states themselves. In other words, dW and dQ, though they depend only on the independent variables which define the state of the system, and on their differentials, will not generally be perfect differentials of functions of the independent variables. This is one of the most important results in the whole science of energy.

When the only forces which do work between two bodies A, B are contact-forces, let the total work done by A on B consist of a quantity of mechanical work W_1 and a quantity of non-mechanical work Q_1; also let W_2, Q_2 be the corresponding quantities of work done by B on A. Then we have

$$(W_1 + Q_1) + (W_2 + Q_2) = 0,$$

so that Q_1 and Q_2 cannot be equal and opposite unless W_1 and W_2 be so too. This requires that the surfaces of the

two bodies should not slip over one another. For example, if the surface of A be at rest while that of B slides, we shall have W_1 negative, equal to $-W$, say, and W_2 zero. Hence
$$Q_1 + Q_2 - W = 0,$$
so that $Q_1 + Q_2$ is not zero.

16. In one special case, of frequent occurrence, the value of dW takes a very simple form. If, for example, a body is exposed to the air, it will be acted on by only three external forces which do mechanical work—the pressure of the air, the attraction of the earth, and the reactions of the fluid or solid bodies with which it is in contact. If the centre of mass of the body move neither up nor down, and if the reactions of the contiguous objects do no work, the two latter forces need not be considered. The only force with which we are concerned is therefore a uniform normal pressure, which will generally also be constant, on certain parts of the surface.

The uniform normal pressure on the surface being denoted by p, and the volume by v, the pressure on a

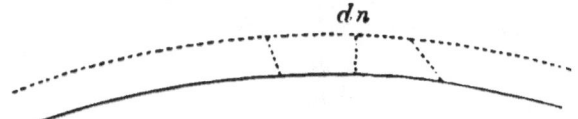

small area da of the surface will be pda, and hence when this small area is forced out a small normal distance dn by the expansion of the body, the mechanical work done upon it by the pressure will be
$$-pdadn.$$
Since the pressure is uniform, the total mechanical work, dW, done on the body in any given short time will be
$$-p\iint dadn,$$

the integration extending over all parts of the surface exposed to the air.

But when the only parts of the surface at liberty to expand are those which are exposed to the normal pressure, $\iint d a d n$ will be the increase in volume, which we denote by dv: hence

$$dW = -pdv \quad \ldots\ldots\ldots\ldots\ldots\ldots(9),$$

and therefore $dU \equiv dW + dQ = dQ - pdv.$

The expression pdv will be a complete differential if p be a function of v only, or a constant, or if v be constant. When p is constant,

$$dW = -d(pv),$$
and $$dQ = d(U + pv).$$

The mechanical work done by the body during a change of volume may be represented graphically by means of a diagram, first employed by Watt for the steam engine and often known as Watt's Diagram of Energy, or an Indicator diagram. Two rectangular axes being taken,

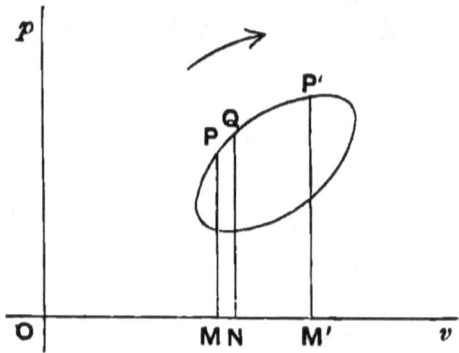

the volume of the body and the uniform normal pressure to which it is subjected at any instant are represented by the abscissa and ordinate of a point P in the plane of the axes. If, then, P, Q are two consecutive positions of the point P, and PM, QN perpendiculars on the axis of volumes, the mechanical work done by the body in the corresponding small change of state, which, as we have already seen, is equal to pdv, will be proportional to the elementary area $PMNQ$. Hence if PP' denote a finite change of state, the corresponding mechanical work done by the body will be proportional to the area of the figure $PMM'P'$, and will evidently depend on the form of the curve joining PP' as well as on the positions of P and P' themselves. If, however, the normal pressure be constant, PP' will be a straight line and the area $PMM'P'$ will depend only on the positions of P and P'. If the volume be constant, the area $PMM'P'$ will always be zero.

17. We are now able to explain a practical method of measuring non-mechanical work, depending on the fact that when water at $0°$ C. is subjected to a constant normal pressure of one atmo, it is able to take up a state of mechanical rest either in the solid or in the liquid form.

In the first place, it appears from observation and may also be proved by calculation, that the force of gravitation between small bodies is exceedingly minute, and therefore the work which it does will be insignificant—since it depends only on the mechanical, or visible, motions. Hence if we have a number of small bodies near the surface of the earth, whose centres of mass move neither up nor down, we need neither consider the attraction of the earth nor their attraction on one another. Also by wrapping the bodies well up in the best non-conducting materials, we may approximate very closely to an ideal case in which there is neither conduction nor radiation of heat from the bodies; that is, no non-mechanical work.

34 ELEMENTARY THERMODYNAMICS.

Consequently, if these bodies undergo an operation, the only forces with which we have to deal will be contact-forces which do no non-mechanical work. Suppose, then, that we take such a system consisting of a quantity of ice and two bodies A, B, exposed only to the following contact-forces:—

(1) Controllable contact-forces acting on A and B.

(2) A constant normal pressure of one atmo over all parts of the surface of the ice which are at liberty to expand.

The system being originally in a state of mechanical rest at $0°$ C., let a measured quantity of mechanical work, W ergs, be expended in rubbing the two bodies A, B together, and after allowing a sufficient time for the system to come to a state of mechanical rest at the same temperature as before, suppose the only effect produced to be the conversion of n grammes of ice into water. Then it is clear that $\frac{W}{n}$ is the quantity of non-mechanical work, in ergs, that must be done on one gramme of ice at $0°$ C. to convert it into water at the same temperature under a constant pressure of one atmo. The value of $\frac{W}{n}$ is found to be 3,292,025,964; and since, at a pressure of one atmo, the volume of one gramme of ice at $0°$ C. is 1·087 cubic centimetres and the volume of one gramme of water only 1·00011 cubic centimetres, the work done by the pressure of the air on the ice will be negative and numerically equal to ·08689 × 1013510, that is, 88060, ergs, per gramme of ice melted.

Secondly, the non-mechanical work done on any system

in any change of state, being the excess of the total increase of energy over the mechanical work done on the system, can be found as soon as the increase of energy is known, and this may be determined for conveniently small bodies in the following manner:—

(1) In the initial state, let the body be protected from all external influences until the whole of the mechanical kinetic energy is expended in internal friction or collisions. Then place the body in a vessel surrounded by ice at $0°$ C., protected by the best non-conducting materials and subject to a constant normal pressure of one atmo, and suppose that when an invariable state has been attained at $0°$ C., the only effect on the vessel and the ice which surrounds it is that m_1 grammes of ice have been converted into water at the same temperature. Then, since the radiant energy in the interior of the vessel is far too small to be taken into account, and since it is evident that there is no radiation from the exterior surface of the ice and that the normal pressure does no non-mechanical work, the energy lost from the given body during the cooling process will be

$$m_1 (3{,}292{,}025{,}964 - 88060) \text{ ergs.}$$

(2) In the final state, let the body undergo the same processes as before, and suppose its ultimate state is the same as its ultimate state in (1). Then if m_2 grammes of ice be converted into water during the processes, it is clear that the energy in the final state exceeds the energy in the initial state by

$$(m_2 - m_1)(3{,}292{,}025{,}964 - 88060) \text{ ergs.}$$

18. The increase of the non-mechanical kinetic energy

depends only on the initial and final states, while the non-mechanical work generally depends on the way in which the change of state takes place, as well. The non-mechanical work done on a system is therefore not generally equal to the corresponding increase of the non-mechanical kinetic energy. This also follows from the fact that the non-mechanical kinetic energy of a system may be altered by friction or collision between the different parts of the system, even when there are no external forces. We are consequently unable to apply the word 'heat' indifferently to non-mechanical kinetic energy and to non-mechanical work. Now we have a practical method of determining non-mechanical work; but we are unable to measure non-mechanical kinetic energy, because we have never yet been able to deprive a body of all its non-mechanical motions. We shall therefore always use the word 'heat' as an equivalent for non-mechanical work, whenever we wish to speak with scientific accuracy.

19. It is considered to be proved, by observation and experiment, that grittiness, the cause of friction, is a universal property of matter, and it may be concluded, from the preceding and similar arguments, that it is the sole cause of the existence both of non-mechanical work and of non-mechanical kinetic energy, and that it considerably modifies radiation; while the rigid accuracy with which all known actions at a distance fulfil their fixed laws is supposed to prove that friction is entirely absent in the ether. The general tendency we have already noticed, for bodies to assume an apparently invariable form and internal condition, is evidently a consequence of friction. It will be seen hereafter to be a case of

THE CONSERVATION OF ENERGY.

Carnot's principle, which is merely a great Law of Friction.

20. When the form and internal condition of a body have been rendered constant by friction, the mechanical motions of the body admit of a very simple representation.

Let O be a point fixed in the body, and imagine an ideal sphere of unit radius with O for centre to be carried about with O in such a manner that, at every instant, every point of its surface is moving with a velocity equal and parallel to that of O. Then as O moves about, the line joining O to any point fixed in the ideal sphere always remains parallel to itself. Let the lines joining O to two points P, Q, fixed in the body, meet the sphere, at any instant T, in two non-coincident points A, B; and at the time $T+t$, in A', B'. If the great circle which bisects AA' meet the great circle which bisects BB' in I, we shall have $IA = IA'$ and $IB = IB'$. Hence, since $AB = A'B'$,

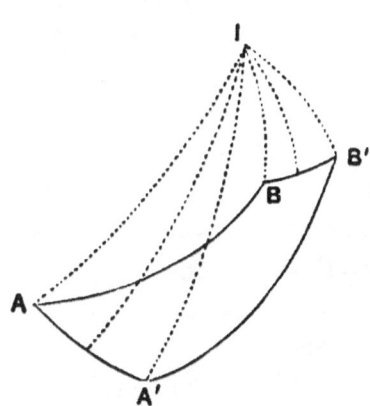

the two triangles IAB, $IA'B'$ have their sides respectively equal to one another, and therefore one is merely the same triangle as the other twisted round OI. Since the whole body may be supposed to be rigidly connected to the three points (O, P, Q), it is evident that the mechanical displacement which has actually taken place might have been effected in either of the following simple ways:—

(1) First, let every part of the body receive a displacement equal and parallel to that of the point O, which will leave the directions of OP and OQ unaltered, and then rotate the whole body about an axis OI, passing through O, until the directions of OP and OQ change from OA, OB to OA', OB', respectively.

(2) Let the body be rotated about an axis passing through O in a direction parallel to OI, through the same angle in the same direction as before, and then give to every part of the body a displacement equal and parallel to that of O.

In order to find the relation between the rotations corresponding to any two different base points O, O', let us suppose the point O to carry about with it three rectangular axes whose directions always remain parallel to themselves, Oz being the axis of rotation through O and the plane zOy passing through the position of the point O' at the time T. If the figure represent the positions of these axes at the time $T + t$, and if Q_1, Q_2 be the positions of O' *relatively* to the axes at the times T, $T + t$, the points Q_1, Q_2 will lie on a circle whose centre is M, the point where the plane through Q_1 and Q_2 parallel to xOy meets Oz. If Q be the original position of O', then when O is chosen for base point, every part of the body receives a displacement equal and parallel to QQ_1, in consequence of which Oz is brought into the position shown in the figure, and afterwards the whole body is rotated about Oz through an angle $\theta \equiv Q_1 M Q_2$. When O' is chosen for base point, every part of the body first receives a displacement

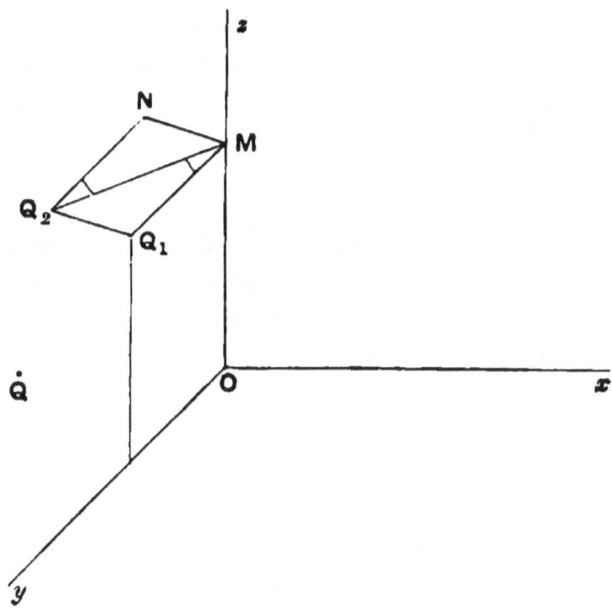

equal and parallel to QQ_2, which is equivalent to a displacement QQ_1 followed by another displacement Q_1Q_2. Hence, when O' is the base point, the axis Oz will be brought by the motion of translation into a parallel position passing through N, where Q_1Q_2MN is a parallelogram. To bring Oz into its final position, we must then rotate the body in the same direction and through the same angle as before about an axis parallel to Oz, passing through Q_2, the final position of O'. Thus the axes of rotation corresponding to all base points are parallel and the angles of rotation equal and in the same direction.

Hence, at any instant, the velocity of any part, P, of

the body is the resultant of the velocity of any base point O, fixed in the body, and the velocity that P would have if O were at rest and the body rotating about an axis OI, passing through O. If a different base point, O', be chosen, the axis of rotation through O' will be parallel to OI and the angular velocity about it equal to that about OI and in the same direction.

For the further discussion of rotations, the principle of Angular Momentum is required.

21. Let Ox, Oy, Oz be any three rectangular axes, and let P be the position of a particle whose mass is m. Suppose the plane through P parallel to xOy to meet Oz in N, and let PV be the direction of the velocity of

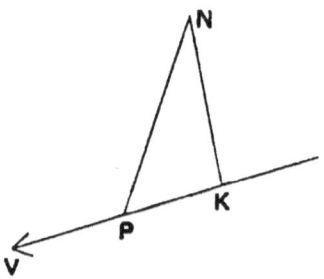

the particle in this plane. Then we define the Angular Momentum, or the Moment of the Momentum, of the particle at P about Oz to be the product of the resolved linear momentum along PV and the perpendicular NK from the point N on PV, the product being reckoned positive when the particle's velocity tends to carry it round Oz in the positive direction and vice versâ. The positive directions round the axes are always taken to be—round Ox, from y to z; round Oy, from z to x; and round Oz, from x to y.

THE CONSERVATION OF ENERGY. 41

The angular momentum round Oz of a finite body, or of a system of bodies, is defined to be the sum of the angular momenta of its ultimate particles about Oz. When a system contains several bodies, its angular momentum about any line will therefore be equal to the sum of the angular momenta of the separate bodies.

The moment of a force about any line Oz is defined in the same way as the angular momentum of a particle, linear momentum being simply replaced by force.

If ϵ be the angle between NK and NP, r the length of NP, v the velocity of the particle along PV, and m its mass, its angular momentum about Oz will be $mv\,(r\cos\epsilon)$.

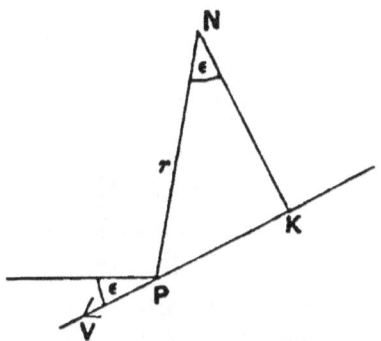

Hence, since $m\,(v\cos\epsilon)$ is the resolved linear momentum perpendicular both to Oz and NP, we see that the angular momentum of the particle about Oz is equal to the product of the perpendicular distance, PN, of the particle from Oz, into the resolved linear momentum at right angles both to Oz and NP, reckoned positive when in the positive direction round Oz.

Again, if PV represent the magnitude of the resolved linear momentum as well as its direction, it is clear that the angular momentum of the particle about Oz will be proportional to the area of the triangle NPV. Now if Pa, Pb, Pc, be the components of the total linear momentum of P in any three directions, according to the

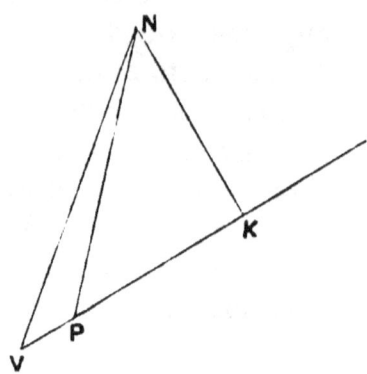

parallelogramic law, the sum of the perpendiculars from (a, b, c) on a plane through NP and Oz will be equal to the perpendicular from V on NP. Hence if (a', b', c') be the projections of (a, b, c) on the plane NPV, the sum of the perpendiculars from (a', b', c') on NP will be equal to that from V. Consequently, the triangle NPV is equal to the sum of the three triangles NPa', NPb', NPc'; or the angular momentum of a particle about any line Oz is equal to the sum of the moments of its component linear momenta. Similarly it may be shown that the moment of a force is equal to the sum of the moments of its components.

Next, let P, P' be the positions of the particle at two different times t, $t + \tau$, where τ is indefinitely small. Let PV represent the total linear momentum at the time t, and suppose the particle to be acted on at that instant by a resultant force F in the direction PA. Also let the three components of the total linear momentum at the

THE CONSERVATION OF ENERGY.

time $t + \tau$, according to the parallelogramic law, be represented by:—

(1) $P'V'$, equal and parallel to PV.

(2) $P'A'$, representing $F\tau$, in a direction parallel to PA.

(3) $P'B'$, the magnitude and direction of which depend on the way in which the force which acts on the particle varies as the particle moves from P to P'.

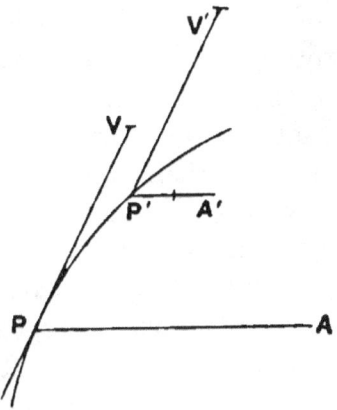

Then if q, q' be the angular momenta of the particle about any line Oz at the times t, $t + \tau$, we have

q' = moment of $P'V'$ + moment of $P'A'$ + moment of $P'B'$,

q = moment of PV.

But $P'B'$ and the distance between PV and $P'V'$ are quantities of the second order: hence if we retain only quantities of the first order, we get

$$q' - q = \text{moment of } P'A'.$$

Now the moment of $P'A'$ may be taken to be the same as the moment of $F\tau$ acting at P in the direction PA, because their difference is a small quantity of the second order: hence $\dfrac{q'-q}{\tau}$ = moment about Oz of the force F acting at P. Proceeding to the limit, we see that, at any instant, the rate at which the angular momentum of the particle about any line is increasing with the time is equal to the moment of the resultant force which acts upon it at that instant. It follows therefore that the rate at which the angular momentum about any straight line of a finite body, or of a system of bodies, increases with the time, is equal to the sum of the moments about this straight line of all the forces, both external and internal, which act upon it. But, *according to the third law of motion*, the internal forces consist of a set of equal and opposite reactions, and consequently the sum of their moments is zero. Hence the rate at which the angular momentum increases with the time is simply equal to the moment of the external forces. If, then, the moment of the external forces about any straight line is constantly zero, the angular momentum of the system about this straight line will remain constant.

22. Suppose the annexed figure to represent the projections of the various lines on a plane through G, the centre of mass, parallel to xOy. Let QR be the projection of the direction of the velocity of a particle P whose mass is m. Draw OQ and GR perpendicular to QR, and let a line through G parallel to RQ meet OQ in S. Then if v be the velocity of the particle along QR, its angular momentum about Oz will be $mv \cdot OQ$, or $mv \cdot GR + mv \cdot OS$. The term $mv \cdot GR$ is the angular momentum of the particle about an axis through G parallel to Oz: the other

term, $mv \cdot OS$, is the angular momentum about Oz of a particle of mass m situated at G and moving with a velocity equal and parallel to that of P. The angular

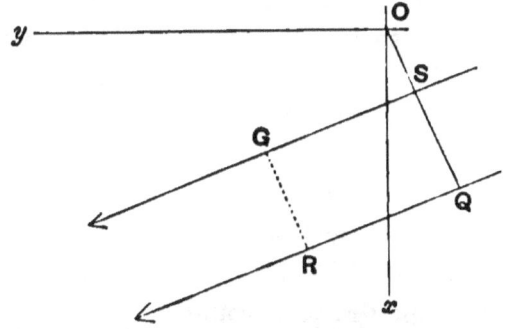

momentum of the whole system about Oz is therefore equal to the angular momentum about a parallel axis through G, together with what the angular momentum about Oz would be if all the particles were transferred to G without altering their velocities either in magnitude or direction. But, since G is the centre of mass, these ideal particles at G are known to be equivalent, as to linear momentum, to a single particle of mass M situated at G and moving about with it, M being the mass of the whole system. Hence, since the angular momentum of a particle is equal to the sum of the moments of its component linear momenta, it follows that the angular momentum of the system about Oz is equal to the angular momentum about a parallel axis through G, together with the angular momentum about Oz of a mass M placed at G and moving about with it. We shall refer to these two parts as the angular momenta due to rotation and translation, respectively.

In like manner it may be shown that the moment of a

force about Oz is equal to its moment about a parallel axis through G (or any other point G'), together with the moment about Oz of an equal and parallel force acting at G (or G'). Thus the moment of all the external forces about Oz is equal to their moment about a parallel axis through G (or G'), together with the moment about Oz of the resultant of a system of equal and parallel forces acting at G (or G').

As the body moves about, let G, G', be the positions in space of the centre of mass at any two consecutive instants t, $t+\tau$; and let GN, $G'N'$ be the corresponding positions of the straight line drawn through G parallel to the fixed line Oz. Also let q be the angular momentum of the body about GN at the time t, and q' the angular momentum about $G'N'$ at the time $t+\tau$. Then since the angular momentum of the body about GN at the time $t+\tau$ is equal to the sum of q' and of the moment of a mass M situated at G and moving about with it, it is evident that the angular momentum about GN at the time $t+\tau$ differs from q' only by a small quantity of the second order, and that the rate at which the angular momentum about a fixed straight line, coinciding with GN, is increasing with the time at the time t, is equal to the limit of $\dfrac{q'-q}{\tau}$. We therefore see that at any instant, the moment of the external forces about the moving axis through G parallel to Oz is equal to the rate at which the angular momentum about that axis is then increasing with the time, that is, is equal to the rate of increase of the angular momentum of rotation. From this it follows that the moment about Oz of a system of forces applied at G, equal and parallel to the external forces, must be equal to the rate of increase of the angular momentum of translation. Hence, when there are no external forces, the angular momenta, both of translation and rotation, remain constant.

Again, the velocity of any particle P may be supposed to consist of a velocity equal and parallel to that of G, the centre of mass, combined with the velocity of P relative to G. If p be the distance of P from a plane through the

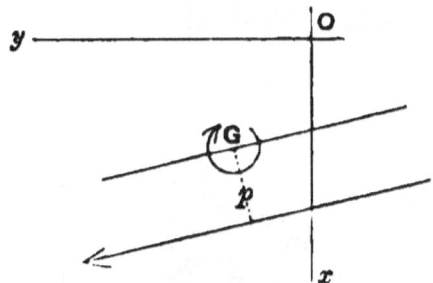

direction of G's velocity perpendicular to xOy, the angular momentum due to the former component about an axis through G parallel to Oz will be mpV, where V is the velocity of G parallel to the plane xOy. For the whole system, this is $V\Sigma mp$, distances on one side of the plane through G being considered positive, on the other negative. But by a well known property of the centre of mass, $\Sigma mp = 0$. Hence, finally, the angular momentum of rotation depends only on the velocities of the particles relative to the centre of mass.

In consequence of these important properties, we shall always suppose the mechanical motions of a body of which the form and internal condition are sensibly invariable, to consist of the motion of translation of the centre of mass combined with a rotation about an axis passing through the centre of mass.

23. The preceding principles immediately lead to some very valuable results relating to bodies of a permanent internal condition. Thus let a body free from mechanical vibrations be rotating with angular velocity ω about an axis passing through G, the centre of mass. Take three rectangular axes Gx, Gy, Gz, one of which, Gz, coincides with the axis of rotation. Then any small portion of the body will describe a circle about G in a plane parallel to xGy. Hence if r be the distance of any

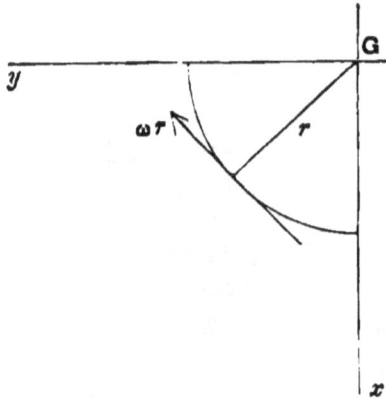

particle P from Gz, its mechanical motion relative to G will simply consist of a velocity ωr at right angles both to Gz and to the perpendicular from P on Gz. Besides this, there may be velocities (α, β, γ) relative to G parallel to the axes, due to the non-mechanical motions. The angular momentum about Gz of the motion of P relative to G, being the sum of the moments of the component linear momenta relative to G, will therefore be

$$mr^2\omega + m(\beta x - \alpha y),$$

where m is the mass of the particle.

The angular momentum of the whole body about Gz is consequently equal to

$$\omega \Sigma m r^2 + \Sigma m (\beta x - \alpha y).$$

But the angular momentum is clearly zero when the body does not rotate, that is, when $\omega = 0$, and the non-mechanical motions are the same as before. Hence

$$\Sigma m (\beta x - \alpha y) = 0,$$

and if C be the 'Moment of Inertia' of the body about Gz, or $C \equiv \Sigma m r^2$, the angular momentum about the axis of rotation takes the very simple form $C\omega$.

It is usual to put $C = Mk^2$, where M is the mass of the whole body. The length k is known as the 'radius of gyration' about Gz, and it is evident that the angular momentum of the body about Gz is the same as that of an ideal particle of mass M situated at a distance k from Gz and revolving round Gz with angular velocity ω.

The angular momenta of the body about Gx and Gy are not necessarily zero, although the body is rotating about an axis at right angles to both of them. In fact, since the velocity ωr is equivalent to $-\omega y$ parallel to Gx and $+\omega x$ parallel to Gy, the angular momentum about Gx is equal to

$$-\omega \Sigma m x z + \Sigma m (\gamma y - \beta z).$$

The second term may be shown to be zero as before, and thus the final result is $-\omega \Sigma m x z$. Similarly the angular momentum about Gy is $-\omega \Sigma m y z$.

If the centre of mass be in motion, as in the case of a planet revolving round the sun, and if Ω be its angular velocity about an axis Oz' parallel to Gz, the angular momentum of translation about Oz' will be $MR^2\Omega$, where

R is the distance of G from Oz'. The total angular momentum about Oz' is therefore

$$MR^2\Omega + C\omega \dots\dots\dots\dots\dots(10).$$

To find the kinetic energy, let (u', v', w') be the velocities of P, (u, v, w) those of G, so that

$$\left. \begin{array}{l} u' = u - \omega y + \alpha \\ v' = v + \omega x + \beta \\ w' = w + \gamma \end{array} \right\}.$$

Then

$$\tfrac{1}{2}\Sigma m\,(u'^2 + v'^2 + w'^2) = \tfrac{1}{2}\,(u^2 + v^2 + w^2)\,\Sigma m$$
$$+ \tfrac{1}{2}\omega^2 \Sigma m\,(x^2 + y^2) + \tfrac{1}{2}\Sigma m\,(\alpha^2 + \beta^2 + \gamma^2).$$
$$- u\Sigma m\alpha + v\Sigma m\beta + w\Sigma m\gamma + \omega\,[-u\Sigma my$$
$$+ v\Sigma mx + \Sigma m\,(\beta x - \alpha y)].$$

But we have

$$\Sigma m\,(\beta x - \alpha y) = 0,$$
$$\Sigma mx = \Sigma my = 0, \text{ since } G \text{ is the origin;}$$

also
$$\Sigma mu' = \Sigma mu, \text{ or } \Sigma m\alpha = 0,$$
$$\Sigma mv' = \Sigma mv, \text{ or } \Sigma m\beta = 0,$$
$$\Sigma mw' = \Sigma mw, \text{ or } \Sigma m\gamma = 0,$$

and
$$\Sigma m = M.$$

Hence we obtain

$$\tfrac{1}{2}\Sigma m\,(u'^2 + v'^2 + w'^2) = \tfrac{1}{2}M\,(u^2 + v^2 + w^2) + \tfrac{1}{2}C\omega^2$$
$$+ \tfrac{1}{2}\Sigma m\,(\alpha^2 + \beta^2 + \gamma^2)\dots\dots\dots\dots(11).$$

The term $\tfrac{1}{2}M\,(u^2 + v^2 + w^2)$ is called the mechanical kinetic energy of translation, being the same as the kinetic energy of a single particle of mass M moving about with G, the centre of mass. The term $\tfrac{1}{2}C\omega^2$, or $\tfrac{1}{2}Mk^2\omega^2$, is called the mechanical kinetic energy of rotation, and is the same as the kinetic energy of a single particle of mass M moving with velocity $k\omega$.

24. Again, if at any instant t, a body be rotating about a straight line Gz, those parts of the body which then lie on Gz will all be moving with velocities equal and parallel to that of G. Hence in any short interval τ, the displacements of these parts will be equal and parallel to that of G, and thus, at the instant $t + \tau$, they will still lie on a line through G parallel to the original direction of Gz. If, therefore, at the time $t + \tau$, the body be rotating about an axis Gz', drawn through G in a different direction from Gz, it is clear that Gz and Gz' cannot have the same position with respect to the body. This fact is usually expressed by saying that when the axis of rotation moves about in space, it also moves about in the body. In like manner it may be shown that when the axis of rotation is fixed, as to direction, in space, it is also fixed in the body.

Now it is evident that the internal mechanical stresses of a body depend only on its form and internal condition, and therefore when these are invariable, the stresses will be constant in magnitude and in directions which are fixed with regard to the body. A little consideration tells us that, when there are no external forces, this condition is inconsistent with a change of the axis of rotation. Thus a body which is under the action of no forces and possesses angular momentum about any straight line through G, must either be rotating about an axis through G fixed in the body and in a constant direction in space or be in a state of mechanical vibration. Now if we assume the tendency of bodies under the action of no force to take an invariable internal condition to be universal, it follows that the mechanical vibrations will all ultimately disappear. But since there are no external forces, the angular momentum of the body about any straight line drawn in a fixed direction through G, the centre of mass, is constant. The mechanical motions of the body relative to G cannot therefore all disappear and we conclude that the body will ultimately settle down into a state of steady rotation about an axis fixed in the body which passes through G in a constant direction.

When a body subject to no external influences has taken up its final state of rotation about an axis Gz, let the point G be imagined to carry about with it three rectangular axes Gx, Gy, Gz, whose directions always remain parallel to themselves. Then since $C\omega$ is the angular momentum about Gz, ω will be constant, and since $-\omega\Sigma mxz$ is the angular momentum about Gx, a line drawn in a fixed direction, Σmxz will also be constant. Now if Gx', Gy' be any two axes at right angles to one another and to Gz which are carried round with the body,

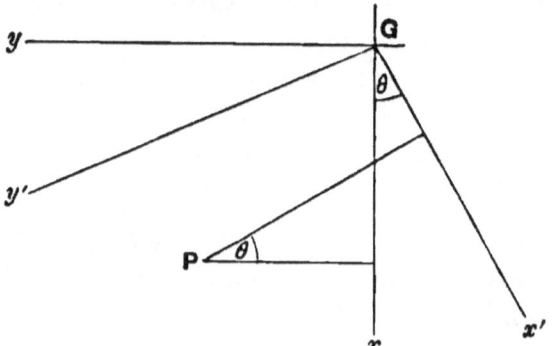

$\Sigma mx'z$ and $\Sigma my'z$ will evidently be constant. Also if θ be the angle between Gx and Gx', we shall have
$$x = x'\cos\theta + y'\sin\theta,$$
and therefore
$$\Sigma mxz = \cos\theta \Sigma mx'z + \sin\theta \Sigma my'z.$$

Since θ will pass through every value as the axes Gx', Gy' revolve round Gz, this result shows that Gz makes

$$\Sigma mzx' = \Sigma mzy' = 0,$$

and
$$\Sigma mzx = \Sigma mzy = 0,$$

or the angular momentum of the body about any line at right angles to the axis of rotation is zero.

25. We can now explain some further results which are of the highest interest and importance in thermodynamics.

First, suppose all the external forces which act on a body to be actions at a distance such that the force on any particle is equal to mg, where at any instant, g is the same in magnitude and direction for every particle of the body. Then, since it is clear that such forces have no effect in producing internal stresses, and since they give no moment about any axis through G, the centre of mass, the body will behave, as to rotations, exactly as if it were under the action of no forces. It will therefore settle down at last into a state of uniform rotation about an axis passing through G in a constant direction. The mechanical kinetic energy of the body will then be

$$\tfrac{1}{2} M (u^2 + v^2 + w^2) + \tfrac{1}{2} C \omega^2,$$

where M is the mass of the body, ω the uniform angular velocity, and (u, v, w) the velocities of G parallel to three rectangular axes Ox, Oy, Oz, fixed in space.

But if g be equivalent to (g_x, g_y, g_z) parallel to these axes, we have

$$m_1 g_x + m_2 g_x + m_3 g_x + \ldots = M \frac{du}{dt},$$

or
$$g_x = \frac{du}{dt}.$$

Again, the work done on the body by the forces parallel to Ox, in the short time dt, will be
$$dW_x = g_x(m_1 dx_1 + m_2 dx_2 + m_3 dx_3 + \ldots).$$
Now if (x, y, z) be the coordinates of G, we have
$$m_1 x_1 + m_2 x_2 + m_3 x_3 + \ldots = Mx,$$
and therefore
$$dW_x = Mg_x\, dx \equiv Mu\frac{du}{dt}\, dt.$$

Thus the total work done on the body by the external forces, in any finite time, is
$$M\int_0^1 \left(u\frac{du}{dt} + v\frac{dv}{dt} + w\frac{dw}{dt}\right) dt \equiv \tfrac{1}{2}M(u_1^2 + v_1^2 + w_1^2) - \tfrac{1}{2}M(u_0^2 + v_0^2 + w_0^2).$$

Hence, since C and ω are both constant, and all the work done by the external forces in this case is mechanical work, we see that the mechanical work done on the body is equal to the increase of its mechanical kinetic energy.

The problem we have just considered is very nearly that of small bodies falling in a vacuum at the surface of the earth; where the magnitude of g is the same both for all particles and for all time, but its direction, though at any instant the same for all particles, is not independent of the time, being carried about with the earth. The problem is also that of the whole earth, roughly speaking, and consequently the mechanical irregularities of its motion, as shown in the tides, &c., are insignificant for so large a body. In the case of the sun, there are disturbing effects of enormous importance due to radiation-forces, by which the whole mass is kept in a state of tumult.

THE CONSERVATION OF ENERGY.

Again, suppose all the external forces which act on a body to be contact-forces of constant magnitude whose positions with respect to any three rectangular axes in fixed directions at G are invariable, the forces being such that they give no moments about these axes and do no non-mechanical work. Then, since there are no radiation-forces, if the body had no angular momentum before the application of the forces, it is evident that, after they are applied, it will ultimately assume an apparently invariable form and internal condition without rotation. When this is the case, the displacement of every part of the surface will be equal and parallel to that of G. Hence if (X, Y, Z) be the total external forces parallel to three fixed rectangular axes Ox, Oy, Oz, the mechanical work which they do on the body in any short time will be

$$Xdx + Ydy + Zdz,$$

where (x, y, z), $(x + dx, y + dy, z + dz)$ are the initial and final positions of G.

Since $X = M \dfrac{du}{dt}$, &c., &c., this is equal to

$$M \left(u \frac{du}{dt} + v \frac{dv}{dt} + w \frac{dw}{dt} \right) dt.$$

Hence in any finite time the mechanical work done on the body will be

$$\tfrac{1}{2} M (u_1^2 + v_1^2 + w_1^2) - \tfrac{1}{2} M (u_0^2 + v_0^2 + w_0^2).$$

Thus there is again no change in the internal condition of the body and the mechanical work done upon it is exactly equal to the increase of its mechanical kinetic energy, while the non-mechanical work is zero.

In both cases, it will be seen that the body is subjected to a non-frictional process in which no non-mechanical work is done upon it and that the only effect produced is a change in the motion of the centre of mass.

In general, when a body is acted on by external forces, the internal condition of the body will not be constant nor the increase of its mechanical kinetic energy equal to the mechanical work done upon it. Suppose, for example, that a body continues to rotate about an axis Gz, drawn through G in a constant direction, under the action of contact-forces whose moment about Gz is N and about the other axes at right angles to Gz zero. Also suppose that the sums of the components of the external forces parallel to the axes are constantly zero and that the point G is at rest. Then if F be the force on any small element of the surface at P in the direction of rotation at right angles both to Gz and PN, the perpendicular from P on

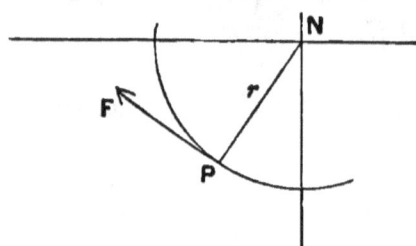

Gz, the mechanical work done by this force in the short time dt will be $Fr\omega dt$, where r is the distance PN and ω the angular velocity of the body. Hence the mechanical work done on the body by all the external forces will be $\omega dt \Sigma Fr$, or $N\omega dt$. Now N is the rate of increase of the angular momentum about Gz, that is, $N = \dfrac{d}{dt}(C\omega)$.

Thus the mechanical work done on the body in a finite time is

$$\int_0^1 \frac{d}{dt}(C\omega)\omega\, dt,$$

or
$$\int_0^1 \left(\omega^2 \frac{dC}{dt} + C\omega \frac{dw}{dt}\right) dt.$$

The increase of the mechanical kinetic energy of the body is

$$\tfrac{1}{2}\int_0^1 \frac{d}{dt}(C\omega^2)\, dt,$$

or
$$\int_0^1 \left(\tfrac{1}{2}\omega^2 \frac{dC}{dt} + C\omega \frac{dw}{dt}\right) dt.$$

These will not necessarily be equal except when $\omega^2 \frac{dC}{dt}$ is constantly zero, which will not generally be true.

If the body be made of very rigid materials, like a fly-wheel, the variations of C will be extremely small, and therefore, if the friction of the axle be neglected, the mechanical work done upon the wheel will be practically equal to the increase of its mechanical kinetic energy.

The rest of the chapter will be devoted to the experimental evidence of the principle of energy. And, for the future, as we shall seldom have any further occasion to make use of the conceptions of 'kinetic energy' and 'work,' in the strict sense of the terms, we shall frequently follow the popular usage by employing these expressions instead of the more correct, but somewhat longer forms, 'mechanical kinetic energy' and 'mechanical work.' Also when we speak about forces, we shall always refer to their mechanical effect: if we wish to refer to the non-mechanical effect, we shall always use the word 'heat.'

26. We have seen that the energy of a material system in any state P is equal to the sum of the kinetic energies of its ultimate particles and of the work which the internal ether-forces do on the system as it returns in any way from the state P to a fixed standard state O. Now since we possess no instruments by means of which the motions of the individual particles can be examined and since gravitation, electric, and magnetic forces are the only ether-forces about which much is known, it is obvious that we have no practical method of obtaining all the details which have been discussed theoretically in the previous part of the chapter. In fact, for experimental purposes, the state of a material system is considered to depend on the following observable particulars only:—

(1) The chemical and physical states and relative positions of the different parts of the system.

(2) Their mechanical motions.

(3) Their temperatures.

(4) Their electric and magnetic states.

On account of the importance of temperature, it is necessary that we should have correct ideas on the subject and that we should understand the method of constructing the mercurial thermometer.

When any system of two bodies A, B is protected from all external influences until the ultimate invariable state is attained, the two bodies A, B are then said to be at the same temperature. In order to test equality of temperatures, we may take a third body, C, as an ordinary mercury thermometer, whose changes of state are easily visible to the eye, and put it in contact, first with A and then with B. If the bodies be protected from all external influences, as before, and C assumes the same final state in both cases, the temperatures of A and B are equal.

If the two bodies A, B, when first placed in contact, do not at once take up an invariable state, their temperatures are not necessarily different. Thus salt and snow may be at the same temperature, as shown by the thermometer, yet when they are mixed together, a violent chemical action takes place by which both are partially melted.

The mercurial thermometer is the instrument most commonly used for detecting differences of temperature. It consists of a closed glass vessel containing only mercury and the vapour of mercury. The glass vessel is in the form of a fine tube, or stem, terminating in a much wider part, called the bulb, and the whole of the bulb and part of the stem is filled with mercury in the liquid state. On heating the thermometer, the glass expands, but the mercury expands more, so that it is forced to rise in the tube, thereby giving a visible effect of change of temperature.

The construction of an accurate thermometer requires five processes, which may be explained as follows.

(1) In the first place we procure a tube of as uniform bore as possible, and then calibrate it, that is, divide it into short lengths of equal capacity. To do this, we force a short column of mercury into the tube and mark the tube at both ends of the column, then move it on its own length until one end comes exactly to where the other was and mark the other end, and so on. After this operation, the bulb A is blown at one end of the tube to a little more than the required size and closed, and, to

enable us to introduce the mercury, a temporary bulb B is blown at the other end and drawn out to a point at which there is a small hole. In forming the bulbs, the air must not be blown by the mouth, for fear of the effects of moisture.

(2) Both bulbs are now gently heated, in order to expel some of the air with which they are filled. The point of B is then immersed in a vessel containing mercury. As the air within the bulbs cools, its expansive

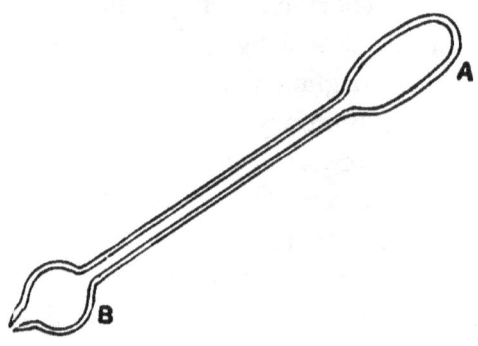

force diminishes, so that a quantity of mercury is forced into the bulb B by the pressure of the atmosphere. The instrument is then set upright, and by alternately heating and cooling the permanent bulb A, a sufficient quantity of mercury is caused to descend into it through the narrow tube from the bulb B.

(3) The instrument is next laid in a sloping position, with the bulb A lowest, on a special furnace and heated till the mercury boils. The vapour of the boiling mercury drives out the air, so then when the boiling ceases, the mercury runs back from the bulb B and completely fills

the tube as well as the bulb A. Then, while the thermometer is still hotter than any temperature at which it is afterwards to be used and the whole of the tube filled with mercury, the top of the tube is melted and closed by a blowpipe flame. As the thermometer cools, the mercury contracts and leaves a space at the top of the tube containing only the vapour of mercury.

(4) The bulb A cannot be blown of exactly the proper capacity for giving the thermometer its required range. For this reason, it is at first made too large and afterwards reduced in the following manner. The bulb A is heated to a temperature shown by a thermometer C, which is already finished, and the point m to which the mercury rises in the stem of A is observed. Next, the bulb A is heated to a different temperature, shown by the thermometer C, and the point n to which the mercury rises in the stem is again observed. Then, knowing the difference of the temperatures shown by C, and the length mn, we can easily calculate how much the bulb A is too large. Lastly, the bulb A is broken off to the required size and closed up again without admitting air.

When the thermometer has been made in the manner just described, it is found that, if left to itself, the bulb undergoes a gradual contraction which, for practical purposes, may be considered complete in six months. We ought, therefore, to keep the thermometer at least six months before proceeding to the last operation.

(5) In graduating the thermometer, we choose two

standard temperatures which are easily reproduced and can be proved by elaborate experimental methods to remain exactly constant. These are the ultimate temperature assumed by a mixture of ice and water when the only external influence is a constant pressure of one atmo, and the temperature of the steam which issues from boiling water at the pressure of an atmo.

In the first place, when the pressure of the air is one atmo, the thermometer is immersed in a mixture of ice and water until the height of the mercury in the tube has become stationary; and the point to which it rises is marked by a scratch on the glass, known as the freezing point. Then the thermometer is surrounded by the steam of boiling water until the mercury is again stationary, and the height to which it rises in the tube is marked by another scratch, called the boiling point. In the Centigrade thermometer, which is the only mercury thermometer now used for scientific purposes, these two temperatures are marked 0 and 100. The portion of the tube between them is then divided into 100 parts of equal capacity, each of which is called a degree, and the divisions are continued beyond the two standard points as far as may be required.

It is necessary that all thermometers should give the same indications of temperature, not merely at the freezing and boiling points, but at all other temperatures. This condition is satisfied, at least approximately, by making all thermometers of the same kind of glass, as will appear from the following argument.

If we take a given mass of mercury, or of glass which is not very irregular in form, and subject it to no stresses greater than the pressure of the air, and if the volume

be v_0 when a standard mercury thermometer is at 0 and v when the standard thermometer indicates θ, it is found by experiment that
$$v = v_0 (1 + \alpha\theta),$$
where α is a constant number, which, for mercury, is about ·00018, and for the glass of thermometers, ·000025. Hence, if V be the volume of the mercury and s the capacity of each degree of the stem when the standard thermometer is at zero, and if the mercury rises n degrees in the stem while it rises θ in the standard thermometer, the volume of the mercury at the temperature θ may be written in either of the forms
$$V(1 + m\theta),$$
or
$$(V + ns)(1 + g\theta),$$
where $m = ·00018$, and $g = 000025$.
We have, therefore, very approximately,
$$V(m - g)\theta = ns.$$
But both thermometers are marked 100 at the boiling point: hence
$$V(m - g) = s,$$
and therefore
$$\theta = n,$$
so that the two thermometers agree.

On account of the difficulty of obtaining glass perfectly homogeneous and of exactly a certain kind, and also on account of the irregular form of the thermometer and the internal strains which are produced in making it, it is found that even when two thermometers have been made in the elaborate manner described above, their readings do not generally quite agree at temperatures distant from the boiling and freezing points. If a thermometer is wanted for very delicate observations, it should therefore be compared with a standard thermometer.

Two other systems of marking the thermometer are often used for non-scientific purposes. In Fahrenheit's thermometer, introduced about 1714, the freezing point is marked 32 and the boiling point 212. In Reaumur's, the freezing point is marked 0 and the boiling point 80. In stating temperatures, it is usual to indicate the scale referred to by the letters C., F., R.

27. It was formerly supposed that both light and heat were material substances. Light was believed to consist of very minute bodies, or 'corpuscles,' which were shot out from luminous bodies with immense rapidity in straight lines. The 'corpuscular theory,' which is principally due to Newton, was sufficient to explain the commoner properties of light. But a different theory, which had been in existence before Newton's time, was again put forward, by Dr Young in England about the beginning of the present century, and in France by Fresnel a little later. Within a few years the new doctrine, according to which light consists of waves of vibrating motion, was so completely proved that it became universally accepted by 1830.

The matter which was supposed to form heat was called 'caloric,' and was regarded as being constant in quantity, like matter in general. According to this view, the conduction of heat was merely the transference of caloric out of one body into another. When a body expanded through rise of temperature, the increase of volume was supposed to be due to the greater amount of caloric present. To explain the increase of temperature that occurs when a body is suddenly compressed, it was natural to suppose that caloric was squeezed out and so rendered sensible to the thermometer or the touch. Again, it was supposed that different bodies required different proportions of caloric to be added to them to produce the same change of temperature: this was expressed briefly by saying that different bodies have different capacities for caloric; thus a pound of water was considered to require 30 times as much caloric to enter it to raise its temperature by 1° as a pound of mercury.

The most important part of the caloric theory, however, was the doctrine of Latent Heat, propounded by Dr Black in 1760. When it was found that by applying heat to a vessel containing ice, no change of temperature was caused so long as the contents of the vessel were kept sufficiently mixed, until the whole of the ice was melted, it was inferred that water merely differed from ice at the same temperature by containing a much larger quantity of caloric. The caloric which it thus appeared necessary to mix with ice in order to change its molecular state without altering its temperature, was called 'latent,' because it could not be detected by the thermometer. In like manner, the steam which issued from boiling water was shown by the thermometer to have the same temperature as the water itself, and was therefore supposed to differ from it only by containing an enormously greater quantity of caloric.

The caloric theory, it must be allowed, afforded a simple explanation of the phenomena just mentioned; but it was not so successful in accounting for the heat developed by friction. It appeared to follow from the theory that, when two bodies were rubbed together, the friction caused a diminution in the capacity of one or both bodies for caloric, in consequence of which the caloric contained within them, without undergoing any increase, was able to raise them to a higher temperature; but it was only necessary to carry on the rubbing process long enough to show that the quantity of heat that could be produced was unlimited, or to test the capacity for caloric of each body after being rubbed, to see that this view was false.

The first person who attained a correct idea of the nature of heat appears to have been Count Rumford. Whilst superintending the boring of cannon in the arsenal at Munich, he was struck with the enormous quantity of heat produced by the working of the steel borer, and in a paper which he published on the subject in the Philosophical Transactions for 1798, he made the remarkable statement that the source of the heat thus developed appeared to be inexhaustible. He then observed that since heat could be furnished by a limited system in unlimited amount, it could not be a material substance. Following up this idea, he argued that heat was motion and made the first attempt to calculate the quantity of mechanical work that must be expended to produce a given quantity of heat.

In the year following, an important experiment was described by Davy. Two pieces of ice were rubbed together until both were nearly melted by the friction, the water thereby produced being a little above the freezing point. Here, then, it was clear that heat was actually created, and, therefore, that it could not be material. Also as the capacity of water for heat was known to be much greater than that of ice, it was evident that friction did not diminish the capacities of bodies for heat in all cases, as the caloric theory required us to believe. Davy was thus led to conclude that 'heat may be defined to be a peculiar motion, probably a vibration, of the corpuscles of bodies.'

Yet notwithstanding these decisive experiments and the progress of the new theory of light, the doctrine of caloric continued to be generally adopted until about 1840. The mechanical theory of heat was then revived almost simultaneously by Mohr, Seguin, and Mayer, who based their ideas on theoretical considerations, and by Colding and Joule, who appealed to experiment. The eloquence of Mayer caused the new theory to be generally assented to: the numerous and brilliant experiments of Joule proved it beyond the possibility of a doubt.

28. In one of Joule's best experiments, water was agitated in a vessel by means of a vertical shaft carrying a number of paddles which worked between fixed vanes, so that when the paddles revolved, the water was prevented from revolving bodily with them, and in consequence, it offered a sufficient resistance. The shaft was caused to rotate by means of a cord wound round its upper part and attached to a heavy weight after passing over a pulley mounted on friction wheels. The water was protected from all external heating and other effects, except the pressure of the air, and the weight could be wound up without moving the paddles. It was then found, after making all corrections, that the work required to be done by the paddles to raise the temperature of the water by any given amount was always proportional to the quantity of water. In addition to the work done by the paddles, there will be a very small amount of work done by the pressure of the air, which, for given initial and final

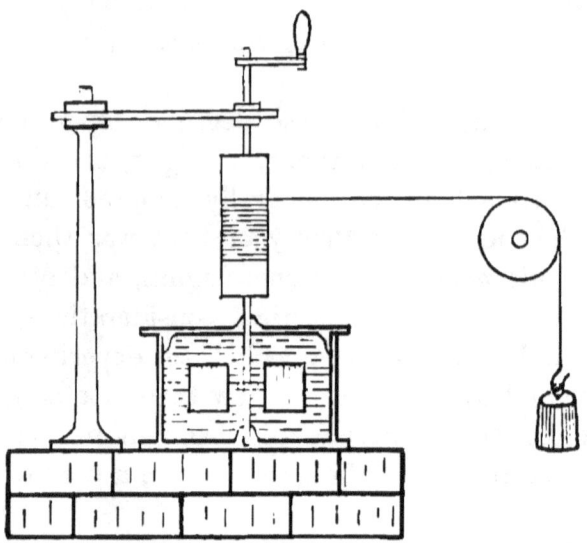

temperatures, will also be proportional to the quantity of water, since, for a short time, the pressure of the air may be considered constant. If, for the sake of numerical illustration, we suppose the mass of water to be a gramme and the pressure of the air exactly an atmo, the work done by the paddles, when the temperature of the water rises from 0° C. to 1° C., is found to be over 40 million ergs; and since the volume of the water at 1° C. is less than the volume at 0° C. by less than the 10,000th of a cubic centimetre, the work done on the water by the pressure of the air will be less than 101·3 ergs.

Experiments were also made by Joule on other substances with a like result. Again, in a number of other experiments, heat was produced indirectly by expending mechanical work on a system so as to cause its parts to become electrified, and then allowing them to resume their unelectrified state; and it was found that the final effect was the same as if the same amount of work had been expended directly. Also it was shown that when heat is developed by chemical action, the ultimate result is the same whether the change takes place by means of electric agency or otherwise.

From these and other experiments it is concluded that, however simple or complicated may be the electric and other properties of any system, or whatever may be the mechanical motions, if it be protected from all external heating and electric effects, the work that must be done on it to bring it from any given state P to any other given state Q depends only on those states and not on the particular manner in which the change of state is produced. To put this statement into a simple form, let us choose a standard state O and denote the work required to bring the system from the state O to the state P by U. Then so long as we keep to the same standard state, U will depend on the state P only and may be written U_P. Also by supposing the system to be brought from the state O to the state Q by first bringing it from O to P and then from P to Q, we see that

$$U_Q - U_P = w_{PQ},$$

where w_{PQ} is the work required to bring the system from P to Q.

Again, if a different standard state O' be chosen and the new value of U be denoted by U', we shall have

$$\left. \begin{array}{l} U'_P = w_{O'O} + U_P \\ U'_Q = w_{O'O} + U_Q \end{array} \right\},$$

so that the value of U is only altered by a constant.

Hence, the quantity U is a single-valued function of the independent variables which define the state of the system, together with an arbitrary additive constant depending only on the choice of a standard state, and dw, the work done on the system in any small change of state, is an exact differential given by $dU = dw$.

The quantity U is called the energy of the system, but it should be noticed that the present definition is not quite the same as that given previously. In fact, the values of U, according to the two definitions, differ by a constant; for the standard state in the former part of the chapter did not make the whole energy zero, but only that part of it which we called potential energy.

If the system consists of a number of bodies which exert forces on one another without being actually in contact, as in the case of gravitation, it is obvious that the energy of the system will depend, not only on the state of each of its parts, but also on their relative positions with respect to one another. That part of the energy which depends on the relative positions of the bodies is called their 'mutual energy,' while the energy of any particular body may be distinguished as its 'internal energy.'

29. Again, if a system be protected from external electric influences only, it is found that no change of state can be produced by the joint agency of heat and work which cannot be produced by work alone, and therefore we may always define the energy of the system in any state P, into which it can be brought by the combined effects of heat and work, to be the work that must be done on the system to bring it from the standard fixed

state O to the given state P on the supposition that both external heating and external electric, influences are absent. Also, since it appears from experiment that in any change of state due to the joint action of heat and work, the increase of the energy is not generally the same as the work actually done on the system, we shall define the heat absorbed by the system in any change of state in which there are no external electric influences, to be the algebraic excess of the increase of the energy over the work done on the system. If the change of state be indefinitely small, and if dU be the increase of energy, dW the work done on the system, and dQ the heat absorbed, we shall therefore have

$$dU = dW + dQ.$$

Again, since a change of state which can be produced by heat alone can also be produced by work alone, or partly by heat and partly by work, in an infinite number of ways, it is evident that in any change of state due to the action of heat and work, the work done on the system and the heat absorbed by it will generally depend on the way in which the change of state takes place as well as on the initial and final states themselves. In mathematical language, the quantities dW and dQ, though functions of the independent variables $(x, y, z,......)$ which define the state of the system, and of their differentials, are not generally complete differentials of functions of $(x, y, z,......)$.

The word 'heat,' in the popular sense, refers to something which exists in a body, and therefore in any change of state, its increase depends only on the initial and final states and not on the way in which the change of state takes place. The popular meaning of the word 'heat' is therefore different from that which we have assigned to it. In fact, according to the popular usage, 'heat' means non-mechanical kinetic energy, while our definition makes it identical with non-mechanical work; non-mechanical work done on the system being the same as heat absorbed, and non-mechanical work done by the system against the external forces, the same as heat given out. To prevent confusion, we shall never use the word 'heat' in the popular sense, but always speak of non-mechanical kinetic energy.

Heat is absorbed either by conduction or radiation. Conduction is a surface phenomenon which takes place when bodies are actually in contact and is merely the transference of energy out of one body into another: radiation is an exchange of energy between a material body and the ether, and may occur in the interior of the body. A common form of conduction is known as convection and may be observed when a pan of cold water is set on the fire; a continual circulation being kept up by the descent of the colder water from above to take the place of the water which ascends after being warmed and made lighter by contact with the heated bottom of the pan. In the case of the sun, an immense flood of light and heat is radiated from the glowing interior through the enormous atmosphere of cooler gases with which it is surrounded.

The principle of the equivalence of heat and work, as expressed by the equation $dU = dW + dQ$, is known as the first Law of Thermodynamics. It is a particular case of the principle of the Conservation of Energy. The second Law of Thermodynamics is Carnot's axiom, which is merely a law of friction.

30. If a system be exposed to external electric influences, it is found that it may acquire such electric properties by contact with other systems that it may be impossible to bring it back to the standard fixed state until the new electric properties are given back. We are then to regard the system as a new system and to choose a new state as the standard state from which to reckon the energy. Again, when the different states of the system are comparable with the same standard state, it is found that, if there are electrified or magnetized bodies in the neighbourhood, the energy of the system may be altered without doing work on it, even when it is evident that there are no external heating influences. The system is then popularly said to absorb 'electric energy,' the meaning of which is fully explained in books on electricity and magnetism. In general, if dU be the increase of energy, dW the work done on the system, and dE the 'electric energy' absorbed, or the 'electric work' done on the system, calculated as in works on electricity, the heat absorbed, dQ, is defined to be the value of $dU - dW - dE$, so that the principle of energy may be written in the form

$$dU = dW + dQ + dE.$$

As we do not wish to enter into the details of electricity and magnetism, we shall, in future, always suppose our system, whether electrified or not, to be so chosen that there are no external electric influences, except in a few special cases. The principle of energy then takes the simple form

$$dU = dW + dQ,$$

where dW and dQ are not generally perfect differentials.

31. The equation
$$dU = dW + dQ$$
may be written
$$dQ = dU - dW.$$
Hence if we suppose the body or system of bodies to undergo a finite change of state, in which U changes from U_0 to U_1, we have
$$\int dQ = U_1 - U_0 - \int dW.$$
If the changes of state be such that the final state is the same as the first, the series of operations which the system goes through is called a *cyclical process*. We have then $U_1 = U_0$, so that
$$\int dQ = - \int dW.$$
Now $-\int dW$ is the external work done by the system: hence when a system which is protected from external electric influences undergoes a cyclical process, the heat absorbed is exactly equivalent to the work done by the system.

32. The result that dW and dQ are not generally perfect differentials is so important that it will be advantageous to consider some simple illustrations.

Suppose, in the first place, that a quantity of water and steam at a high temperature, and therefore also at a high pressure, is contained in a cylinder fitted with a smooth air-tight piston, and let it be required to reduce the contents of the cylinder to a much smaller volume and a lower temperature, say the ordinary freezing point; suppose also that the final volume is too great to be filled by the water within the cylinder, so that steam must also be present and the final pressure inside the cylinder be consequently very small.

The desired change of state may be brought about in many different ways, of which two will be here considered.

(1) Let the piston be forced in so as to reduce the volume of the interior of the cylinder to the required extent before any heat is abstracted, and then reduce the temperature to its final value without altering the volume. The work done on the piston in this method will be considerable.

(2) First suppose heat to be abstracted from the cylinder, without altering the volume, until the desired lower temperature is attained. Then let the piston be forced in while the temperature of the interior of the cylinder is kept constant by the conduction of heat through its sides. The work done in this method will be negligible, since the pressure against which the piston has to be forced is very small.

Since the work done on the system while it experiences a given change of state depends on the way in which the change of state takes place, it follows that dW and dQ are not complete differentials, like dU.

Again, consider the following example. A cylinder fitted with a smooth air-tight piston contains a quantity of air at any given temperature and at volume v, and it is required to increase the volume to v' without altering the temperature.

We will show by two illustrations how this may be done in different ways.

(1) Let the piston be drawn out so slowly that the air within is able to exert its maximum pressure on it and the temperature to be kept constant by the conduction of heat through the sides of the cylinder.

(2) Again, suppose the piston drawn out so rapidly

that the air within is hardly able to keep up with it and therefore unable to do much work, and then let the temperature be brought to its former value.

The work done on the piston by the contents of the cylinder being clearly different in the two cases, we are led to the same conclusions with respect to dW and dQ as before.

33. If a body be subjected to a uniform normal pressure p, the work done on it during a small increase of volume dv is $-pdv$. The equation $dU = dW + dQ$ then becomes $dQ = dU + pdv$. Hence if either p or v be constant, the heat absorbed by the body in any change of state will depend only on the initial and final states. We are accordingly led to the following definitions, in which the state of the body is supposed to depend only on the temperature when either p or v is given—a supposition which requires that there should be neither electric actions nor mechanical motions:—

(1) At any temperature, the 'thermal capacity of a body at constant pressure' is the heat required to raise its temperature one degree C. while the pressure remains constant.

The pressure which most frequently occurs is that of the air, which, at any instant, may be supposed to remain constant for a short time.

(2) At any temperature, the 'thermal capacity of a body at constant volume' is the heat required to raise its temperature one degree C. while the volume remains constant.

If the body is homogeneous, the thermal capacity of a mass of one gramme is called its 'specific heat.'

The specific heat of water at 0° C. under a constant pressure of one atmo is called a Calorie, and is used as an arbitrary unit of heat.

34. In Joule's experiments on the agitation of water, the apparatus cannot be arranged so that the water rises in temperature exactly from 0° C. to 1° C. In order to find the mechanical value of a calorie, we must therefore determine the specific heats of water at different temperatures. This may be done by a method known as the 'method of mixtures.' Thus let a quantity of water at 0° C. be added to an equal quantity at some other temperature, say 100° C., and let the mixture be protected from all external influences except the pressure of the air, which may be taken to be an atmo. Then if the uniform temperature which the water finally assumes be θ C., and if no appreciable amount of work has been done either by gravity or in mixing them together, it follows that, since the work done by the pressure of the air is negligible, the heat required to raise the temperature of any mass of water from 0° C. to θ C., under a constant pressure of one atmo, is equal to that required to raise the temperature of an equal mass of water from θ C. to 100° C. under the same pressure. In this way, the specific heats of water under a constant pressure of one atmo have been determined in calories at different temperatures, thus:—

at 0° C.,......1·000
... 10° C.,......1·0005
... 20° C.,......1·0012
... 30° C.,......1·0020.

By means of these results, it is found from Joule's

experiments that a calorie is equivalent to 42350 gramme-centimetres, or 41,539,759·8 ergs, or about 3 foot-pounds. In English measure, the heat required to raise the temperature of 1 lb. of water from 0° C. to 1° C. under a pressure of one atmo is equivalent to about 1390 foot-pounds.

Now if a mass of one gramme be moving without rotation or vibration at a speed of v centimetres per second, its mechanical kinetic energy will be $\frac{1}{2}v^2$ ergs. If this be equivalent to a calorie, we shall have

$$v^2 = 83{,}079{,}519 \cdot 6,$$
or
$$v = 9114 \cdot 8.$$

Thus the velocity must be 91·148 metres, or 299 feet, per second, that is, 5·469 kilometres, or 3·4 miles, per minute. Hence if two equal masses of water at 0° C. be moving with this velocity and impinge on one another in such a way that they are both brought to rest; then if no steam be formed, the impact will be sufficient to raise the temperature by 1° C. For a rise from 0° C. to 100° C., a velocity of about 55 kilometres, or 34 miles, per minute is required. In the case of iron, the specific heat is only about $\frac{1}{9}$ of a calorie, and therefore the velocities are only about $\frac{1}{3}$ as large as for water.

Again, let x be the latent heat of liquefaction of ice, in calories, under a pressure of one atmo, that is, the number of calories required to convert one gramme of ice at 0° C. into water at the same temperature. Also suppose that i grammes of ice at 0° C. are mixed with w grammes of water at θ C., and that, in consequence, the whole of the ice is melted under a constant pressure of one atmo. Then if no heat be allowed to escape, and if

θ'C. be the final temperature, the heat gained by the water from the ice will be, very approximately, $w(\theta' - \theta)$ calories, and the heat gained by the ice from the water, $i(x + \theta')$ calories. Hence, very approximately,

$$w(\theta' - \theta) + i(x + \theta') = 0,$$

or

$$x = \frac{w\theta}{i} - \frac{w+i}{i}\theta'.$$

It is thus found that the latent heat of ice under a pressure of one atmo is 79·25 calories, or 3,292,025,964 ergs.

Bunsen's calorimeter is a small instrument by which the heat coming from a small solid body under the constant pressure of the air is easily and accurately determined by the melting of ice. It consists of the three parts, a, b, c,

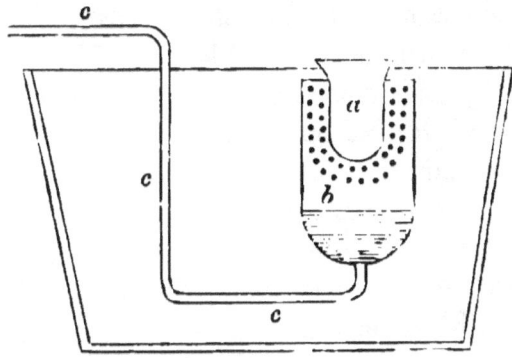

made of glass, and sealed together with the blow-pipe. The part b contains distilled water freed from air by boiling, and the bottom of b and the tube c are filled with boiled mercury, the upper part of the tube being bent horizontally, calibrated, and graduated. In preparing the calorimeter for use, a coating of ice is formed round the test tube a by passing a stream of alcohol, previously cooled below 0° C. in a freezing mixture, down to the bottom of a and back again. The calorimeter is then placed in a vessel filled with clean snow, a substance which soon acquires and long maintains a temperature 0° C., unless the temperature of the room is below 0° C. Lastly, the test tube a is partially filled with water or some other fluid which does not dissolve the body to be experimented on, and as soon as the whole is at the temperature 0° C., the calorimeter is ready for use.

In making an experiment, the body which is to give off the heat is dropped into the test tube a. This will cause the water in a to become warmer, and then, by the conduction of heat through its sides, some of the ice which surrounds it will be melted. This will go on till the temperature of the whole is again reduced to 0° C. If n grammes of ice be melted in the process, the heat given off by the body will be 79·25 calories. The value of n is determined by the movement of the mercury in the graduated tube, depending on the fact that at a pressure of one atmo and at 0° C., one gramme of ice occupies 1·087 cubic centimetres and one gramme of water, only 1·00011. It should be observed that if any air be allowed to remain in the water in b, it will be expelled in the form of a small bubble during the process of freezing the water around the test tube a, and partially re-dissolved when the ice is melted. A small error will thus be introduced into the indications of the calorimeter.

CHAPTER II.

ON PERFECT GASES.

35. We are unable to proceed much further with the first law of thermodynamics until we have introduced Carnot's principle, except in the case of the more permanent gases, where some simple experiments supply us with the information we require. The most common of these gases are Air, Oxygen, Hydrogen, Nitrogen, Carbonic Oxide, Carbonic Acid, Chlorine, Cyanogen, Marsh Gas, Olefiant Gas, Sulphurous Acid, and Ammonia. They are often called 'perfect' gases, because they all exhibit the same simple properties and obey the same laws more or less perfectly; but the first five and Marsh Gas are more perfect than the others.

We shall suppose the gas contained in a closed vessel which is in a state of mechanical rest and either maintained at a uniform temperature or prevented from receiving or losing heat by being wrapped up in some non-conducting material, like felt. For simplicity, we shall also suppose that electric and magnetic actions are entirely absent. Under these conditions it is found that the gas quickly assumes a state of equilibrium in which the temperature has the same value in every part of the vessel. It is also found that the gas exerts a normal pressure against the interior of the containing vessel; and that the pressure of the gas, also the 'density,' that is, the mass per unit volume, and the 'specific volume,' that is, the volume per unit mass, have the same values at all points in the same horizontal plane. Owing to gravity, the density is not quite the same at points not in the same horizontal, but the difference is too small to be taken into account in the present chapter. If gravity be neglected, the density of the gas will have the same value throughout the vessel and the pressure the same value all over the surface. These values may then be referred to briefly as the density and pressure of the gas.

36. A very important experiment on perfect gases, first made by Gay Lussac, was repeated by Joule in 1844 in a greatly improved form. In a vessel of water A, there

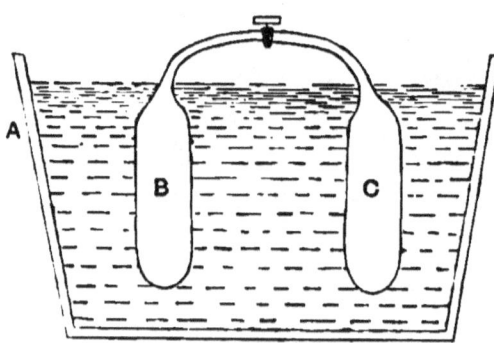

were two strong closed vessels B, C, connected by a pipe in which there was a very perfect stop-cock. Air was compressed to a pressure of about 20 atmospheres in B and exhausted from C, and the temperature of the water in A was made uniform by stirring. The stop-cock was then opened, so that the air rushed from B to C, and, in consequence, the water near B was cooled while that near C was heated. The water in A being again brought to a uniform temperature by stirring and the proper correction made for the work thus expended as well as for the heat lost from A by conduction or radiation, it was found that no perceptible change of temperature had been produced in the water by the expansion of the air. It was therefore inferred that, on the whole, no heat had been gained or lost by the air by means of the water during the operation; and as no external work had been done by it, it followed from the first law of thermodynamics, that the energy was unaffected by the process. In other words, the energy of the air is constant so long as the temperature is constant. Hence we conclude that the energy of a given mass of air in a state of equilibrium is a function of the temperature only.

This remarkable experimental result was tested, both for air and some other gases, by Joule and Thomson a few years later in a series of careful experiments, and was found to be very approximately correct in every case. In consequence, it is generally taken to be true for *all* the more permanent gases.

Hence, if a quantity of any perfect gas expand in any way from one volume to another, and the temperature be made the same after the expansion as before, the energy will be unaltered by the operation, and therefore, by the first law of thermodynamics, the heat absorbed by the gas (in ergs) will be equal to the work done by it during the expansion.

Again, if the temperature of a given mass of gas be

raised by any amount while the volume remains constant, the increase of energy will depend only on the initial and final temperatures, and no work being done by the gas, the same will also be true of the heat absorbed. But the heat absorbed by one gramme of any substance when the temperature is raised 1° C. and the volume kept constant, is the specific heat of the substance at constant volume. In the case of a perfect gas, this is therefore a function of the temperature only, or a constant.

37. There is a remarkable relation between the pressure, density, and temperature of a perfect gas. Suppose, for example, that a gramme of any perfect gas is contained in a cylinder fitted with a smooth air-tight piston, so that the volume of the gas can be increased or decreased at pleasure. Then it is found that, so long as the temperature of the gas is kept constant, the pressure varies inversely as the volume. The pressure of the gas at any instant being denoted by p and the volume by v, it follows that the product pv depends only on the temperature. If we denote the temperature, as indicated by the mercury centigrade thermometer by θ', we may therefore write

$$pv = f(\theta')\quad\quad\quad\quad\quad(12).$$

Again, if the pressure be kept constantly equal to any value p, it is found that the changes of volume due to changes of temperature, may be expressed by the formula

$$v = V(1 + \alpha\theta')\quad\quad\quad\quad(13),$$

where V is the volume when $\theta' = 0$ and α has the same value ·003665, or $\frac{1}{273}$, not only for all pressures, but for all gases.

The former of these laws was discovered by Boyle, the latter by Charles. By continental writers they are generally referred to as the laws of Mariotte and Gay Lussac. It has been shown by Regnault that they are not strictly accurate, but the deviations appear to be very small when the gas is sufficiently removed from its point of condensation, that is, when the pressure is not too great or the temperature too low.

From equation (12) we obtain

$$\frac{v}{V} = \frac{f(\theta')}{f(0)},$$

and therefore, by equation (13),

$$f(\theta') = (1 + \alpha\theta')f(0),$$

that is,
$$f(\theta') = k(1 + \alpha\theta')$$
$$= \frac{k}{273}(273 + \theta'),$$

where k is a constant depending only on the nature of the gas.

If we write θ for $273 + \theta'$ and R for $\frac{k}{273}$, we get

$$pv = R\theta \dots\dots\dots\dots\dots(14),$$

and therefore if (p_0, v_0, θ_0) be any corresponding values of (p, v, θ),

$$\frac{pv}{\theta} = \frac{p_0 v_0}{\theta_0} \dots\dots\dots\dots\dots(15).$$

The result $pv = R\theta$ may be shown, by Carnot's principle, to involve the experimental fact that the energy depends only on the temperature.

38. The simple relation expressed by the equation $pv = R\theta$ for any perfect gas has led to the construction of thermometers in which air is employed as the thermometric substance instead of mercury. In these thermometers, either the pressure or the volume is kept constant. In both kinds, θ is taken to be 273 at 0° C., and R is determined from the equation $pv = R\theta$ by observing the volume at 0° C. corresponding to the given pressure, or the pressure at 0° C. corresponding to the given volume. At any other temperature, the temperature of the air thermometer is defined to be the value of $\frac{pv}{R}$, where, in constant pressure thermometers, the value of v is to be found by experiment, and in constant volume thermometers, the value of p. The indications of constant pressure and constant volume thermometers agree with one another, but they are not quite the same as those of the common mercury thermometer increased by 273. Since any of the perfect gases may be used instead of air, the scale of the air thermometer is often, but improperly, called 'absolute.' A truly 'absolute' scale of temperature, independent of the special properties of any particular substance or class of substances, will be obtained in the next chapter by means of Carnot's principle, and it will be found to coincide practically with the scale of the air thermometer.

In a very simple form of the air thermometer, due to Jolly, the volume of the air is not kept quite constant, but the only change in it is that due to the small expansion by heat of the vessel in which it is contained. This thermometer consists of a glass globe of about 50 cubic centimetres capacity formed in one piece with a fine capillary bent tube A and a larger tube B. The glass globe is filled with dry air and the tube B is joined to an open glass tube C by an india-rubber tube D; the

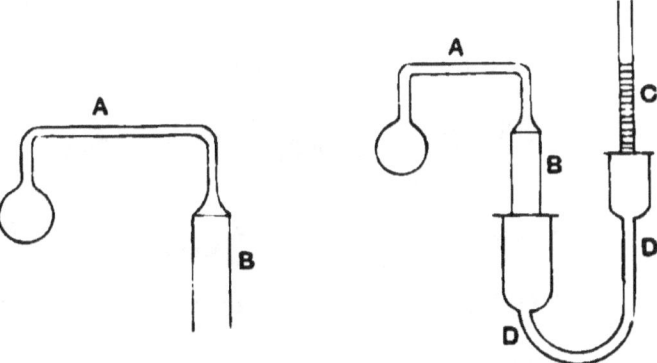

tubes B, D, and the lower part of C being filled with mercury.

In graduating the instrument, we proceed as follows:—

(1) The glass globe is surrounded with melting ice to bring its temperature to $0°$ C. The tube C is then raised or lowered till the surface of the mercury in B is brought to a mark near the capillary tube A. Lastly, the pressure p_0, in the glass globe, is given by the height of the barometer and the height of the mercury in C above the mark on B.

This operation is equivalent to finding the value of R.

(2) The glass globe is then brought to any other temperature we wish to determine, the mercury adjusted and the pressure p obtained, as before.

If V be the capacity of the glass globe at $0°$ C., its capacity when the temperature of the air thermometer is θ, will be
$$V\{1 + \cdot000025\,(\theta - 273)\}.$$
Hence since the quantity of air contained in the capillary tube is negligible, we shall have
$$pV\{1 + \cdot000025\,(\theta - 273)\} = R\theta,$$

and
$$p_0 V = 273 R.$$

Thus
$$\frac{\theta}{273} = \frac{p}{p_0}\{1 + \cdot 000025 (\theta - 273)\},$$

or
$$\frac{\theta - 273}{273} = \frac{p - p_0}{p_0} + \cdot 000025 \frac{p}{p_0} (\theta - 273),$$

whence
$$\theta - 273 = \frac{p - p_0}{\cdot 003665 p_0 - \cdot 000025 p},$$

by means of which θ is found from the observed value of p.

39. The following table[1] exhibits some important fundamental experimental results relating to the perfect gases, at a pressure of one atmo and at 0° C.

	Relative Densities.	Relative Specific Volumes.	Mass of a Litre in Grammes.	Volume of a Gramme in Litres.	Volume of a Pound in Cubic Feet.
Air	1	1	1·2932	·7733	12·39
Oxygen (O)	1·10563	·90446	1·4298	·6994	11·20
Hydrogen (H)	·06926	14·4383	·08957	11·16445	178·85
Nitrogen (N)	·97135	1·02945	1·25615	·7961	12·75
Carbonic Oxide (CO)	·9545	1·0476	1·2344	·8101	12·97
Carbonic Acid (CO_2)	1·52907	·6540	1·9774	·5057	8·10
Chlorine (Cl)	2·4222	·4128	3·1328	·3192	5·11
Cyanogen (NC_2)	1·8019	·5550	2·3302	·4291	6·87
Marsh Gas (CH_4)	·562	1·779	·727	1·375	22·04
Olefiant Gas (C_2H_4)	·982	1·018	1·270	·787	12·61
Ammonia (NH_3)	·5952	1·6801	·7697	1·2992	20·81

By means of these data, we have deduced the values of $\frac{p_0 v_0}{273}$, where v_0 is the volume in cubic centimetres of one gramme of gas at 0° C. and a pressure of one atmo, and (1) p_0 is the number of dynes per square centimetre in a

[1] Everett's 'Units and Physical Constants.'

ON PERFECT GASES.   89

pressure of one atmo, (2) p_0 is the number of gramme-weights (at Paris) per square centimetre in a pressure of one atmo.

	(1)	(2)
Air	2,871,000	2927
Oxygen	2,596,000	2647
Hydrogen	41,448,000	42256
Nitrogen	2,955,000	3013
Carbonic Oxide	3,007,000	3066
Carbonic Acid	1,877,000	1914
Chlorine	1,185,000	1208
Cyanogen	1,593,000	1624
Marsh Gas	5,105,000	5205
Olefiant Gas	2,922,000	2980
Ammonia	4,820,000	4917

40. As an illustration of the first law of thermodynamics in its application to perfect gases, let us suppose a cylinder fitted with an air-tight piston to contain a mass m of any kind of gas at temperature Θ, as measured by the air thermometer, and let the volume be altered so slowly that at every instant the gas is in a state of equilibrium and its temperature kept equal to Θ by the conduction of heat through the sides of the cylinder. Then the force exerted by the gas on the piston will be uniform and its value p, per unit of area, when the volume is v, will be given by the formula

$$pv = mR\Theta,$$

so that the state of the gas will be represented on an indicator diagram by a rectangular hyperbola.

The work, pdv, done by the gas on the piston during a small increase of volume may therefore be written

$$mR\Theta \frac{dv}{v}.$$

Hence when the volume changes from v_1 to v_2, the work done by the gas will be

$$mR\Theta \log \frac{v_2}{v_1}.$$

From this it is clear that when the volumes form a geometrical progression, the corresponding quantities of work will form an arithmetical progression.

Again, if p_1 be the initial pressure, the work done by the gas becomes

$$p_1 v_1 \log \frac{v_2}{v_1}.$$

Hence if two cylinders of the same volume contain any kinds of perfect gas at different temperatures and in such quantities that the pressure is the same in both, each gas will do the same amount of work in expanding by the same amount at constant temperature.

41. A much more important example is obtained by supposing the cylinder to contain one gramme of gas, the temperature of which is slowly raised one degree centigrade (by the air or mercury thermometer) by imparting heat to it under two different conditions:—

(1) Let the temperature be slowly raised while the volume is kept constant.

The increase of energy in this case is equal to the 'specific heat of the gas at constant volume,' and, as we have already seen, is a function of the temperature alone, or a constant. It is usually denoted by C_v (in ergs).

(2) Let the temperature and volume of the gas be slowly increased in such a way that the pressure remains constant.

The heat absorbed by the gas during the process is the 'specific heat of the gas at constant pressure' and is generally written C_p: the increase of energy is the same as before, since the energy of a perfect gas in a state of equilibrium depends only on the temperature. Hence if W be the work done by the gas, we have

$$C_v = C_p - W.$$

But if p be the constant pressure, v and v' the initial and final volumes, we have $W = p(v' - v)$, which is equal to the constant R, by the relation $pv = R\theta$. Thus

$$C_v = C_p - R,$$

or $$C_p - C_v = R \dots\dots\dots\dots\dots(16).$$

This simple relation, in the hands of Clausius and Rankine, furnished some of the earliest triumphs of the mechanical theory of heat.

Prior to 1850, it was supposed to have been established by experiment that the specific heat of a perfect gas depended on the pressure, but in that year Clausius was led by the new theory of heat to assert that the specific heat, whether at constant pressure or at constant volume, could depend only on the temperature, and he conjectured that it would be found to be constant. The fact that the new theory disagreed with the results of accepted experiments led to an attack upon it by Holtzmann. Three years later, however, Regnault published his splendid experiments on the specific heats of gases, by which the conclusions drawn from the mechanical theory were decisively shown to be true; the specific heat of a permanent gas being found to be independent both of the pressure and the temperature.

A still more important application of equation (16) was made by Rankine a little later in the same year 1850, as follows:

The velocity of sound in air depends on the relation between its pressure and density during the rapid condensations and rarefactions as the sound travels along. These changes of pressure and density occur at least hundreds of times in a second, and consequently the heat developed by compression has not time to get away by conduction before the air is again in its natural state as to temperature and density. A formula is thence obtained for the velocity of sound in air which contains the ratio of the two specific heats, $\dfrac{C_p}{C_v}$, usually written k. Comparing the formula with the velocity of sound in air, as observed by Bravais and Martens, it is found that

$$\frac{C_p}{C_v} \equiv k = 1\cdot 410.$$

Hence, by equation (16), we have

$$1 - \frac{1}{k} = \frac{R}{C_p},$$

or $\quad C_p = \dfrac{k}{k-1} R = \dfrac{1\cdot 41 \times 2{,}871{,}000}{\cdot 41}$

$$= 9{,}873{,}000,$$

and $\quad C_v = 7{,}002{,}000.$

If c_p, c_v, be the specific heats in calories, we have

$$c_p = \frac{9{,}871{,}000}{41{,}539{,}759\cdot 8} = \cdot 2377,$$

and $\quad c_v = \cdot 1686.$

The value previously accepted for c_p at atmospheric

pressure was ·2669. Regnault's experimental result, obtained a few years later, was $c_p = ·2375$.

42. There is another method of estimating the specific heats of gases, first calculated by Clausius, which is often useful; viz., the ratio of the heat required to raise the temperature of a quantity of gas at constant pressure or volume to the heat required, under the same conditions as to pressure or volume, to raise the temperature of an equal volume of air to the same extent. These two specific heats will be denoted by γ_p and γ_v, respectively.

Since we have
$$C_p - C_v = R,$$
we obtain, for air,
$$c_p - c_v = \frac{2927}{42350} = ·0691.$$

If v' be the volume of one gramme of another gas at the same temperature θ and pressure p as air, and if R' be the corresponding value of R, we have
$$\left. \begin{array}{l} pv' = R'\theta \\ pv = R\theta \end{array} \right\}.$$

Hence
$$R' = R\frac{v'}{v} \equiv Rx, \text{ say,}$$

where x is the relative specific volume of the gas compared with air, as given in the table.

We have, therefore, for any gas
$$C_p' - C_v' = R' = Rx,$$
or
$$c_p' - c_v' = ·0691x \quad \text{................(17)},$$
a formula which enables us easily to calculate c_v' from Regnault's experimental determination of c_p'.

Again, the quantity of heat which unit volume of

the gas absorbs when its temperature is slowly raised at constant pressure by 1° C. is $\dfrac{C_p'}{v'}$. Hence

$$\gamma_p' = \dfrac{C_p'}{v'} \div \dfrac{C_p}{v} = \dfrac{C_p'}{C_p} \cdot \dfrac{v}{v'} = \dfrac{c_p'}{c_p} y,$$

where y is the density of the gas compared with air, as in the table.

Substituting for c_p its value ·2375, as found by Regnault, we get

$$\gamma_p' = \dfrac{c_p' y}{·2375} \quad\ldots\ldots\ldots(18).$$

In like manner, we have,

$$\gamma_v' = \dfrac{C_v'}{C_v} y = \dfrac{c_v'}{c_v} y = \dfrac{(c_p' - ·0691x)\, y}{c_v}$$

$$= \left(\dfrac{c_p'}{c_p} y - \dfrac{·0691}{c_p}\right) \dfrac{c_p}{c_v} = (\gamma_p' - ·2909) \dfrac{c_p}{c_v} \ \ldots\ldots(19).$$

By means of these formulæ, the accompanying table is calculated from Regnault's experimental determinations of the specific heats at constant pressure in calories.

	Specific heat at constant pressure.		Specific heat at constant volume.		k
	In calories.	Compared with an equal volume of air.	In calories.	Compared with an equal volume of air.	
Air	·2375	1	·1684	1	1·410
Oxygen (O)	·21751	1·012	·15501	1·018	1·403
Hydrogen (H)	3·40900	·994	2·4114	·992	1·414
Nitrogen (N)	·24380	·997	·17266	·996	1·412
Carbonic Oxide (CO)	·2450	·985	·1728	·978	1·418
Carbonic Acid (CO_2)	·2169	1·396	·1717	1·559	1·263
Chlorine (Cl)	·12099	1·234	·0925	1·330	1·308
Marsh Gas (CH_4)	·5929	1·403	·4700	1·568	1·260
Olefiant Gas (C_2H_4)	·4040	1·670	·3337	1·946	1·211
Ammonia (NH_3)	·5084	1·274	·3923	1·386	1·296

43. If we consider a gramme of perfect gas in a state of equilibrium, the three quantities (p, v, θ) satisfy the relation $pv = R\theta$ and therefore any two of them may be chosen as independent variables. A confusion then arises as to the meaning of a partial differential coefficient. Thus $\dfrac{dU}{d\theta}$ will have different meanings according as θ and v or θ and p are the independent variables. To remove this difficulty, Clausius has introduced a very convenient method of notation which has been generally adopted. He denotes the partial differential coefficient of U with respect to θ when θ and v are independent variables by $\dfrac{d_v U}{d\theta}$. Similarly, the partial differential coefficient of U with respect to θ when θ and p are the independent variables is written $\dfrac{d_p U}{d\theta}$.

With this notation, we have, by Taylor's theorem,

$$dU = \frac{d_v U}{d\theta} d\theta + \frac{d_\theta U}{dv} dv,$$

or, since $\dfrac{d_\theta U}{dv} = 0$ for perfect gases, and $\dfrac{d_v U}{d\theta} = C_v$,

$$dU = C_v d\theta,$$

and therefore
$$U = C_v \theta + C' \quad \ldots \ldots \ldots \ldots \ldots \ldots (20),$$

where C' is an arbitrary constant depending on the choice of a standard state.

If the volume and temperature of the gas alter so slowly that at every instant the gas is in a state of equilibrium, the force by which the gas is compressed will only be just sufficient to overcome the pressure of the gas. Hence
$$dW = -pdv,$$

and therefore the equation
$$dU = dW + dQ$$
becomes
$$dU = dQ - pdv,$$
whence
$$\begin{aligned}dQ &= dU + pdv \\ &= C_v d\theta + pdv \\ &= C_v d\theta + \frac{R\theta}{v} dv \end{aligned} \quad \ldots\ldots\ldots(21).$$

If θ and p be taken for independent variables,
$$\begin{aligned}dQ &= C_v d\theta + pdv \\ &= C_v d\theta + \{d(pv) - vdp\} \\ &= (C_v + R) d\theta - vdp \\ &= C_p d\theta - vdp \end{aligned} \quad \ldots\ldots\ldots(22);$$

or thus:—
$$\begin{aligned}dQ &= C_v d\theta + pdv \\ &= C_v d\theta + p\left(\frac{d_p v}{d\theta} d\theta + \frac{d_\theta v}{dp} dp\right) \\ &= (C_v + R) d\theta - \frac{R\theta}{p} dp, \text{ since } pv = R\theta, \\ &= C_p d\theta - \frac{R\theta}{p} dp.\end{aligned}$$

If we wish to take p and v as the independent variables, we may write $\frac{1}{R}(pdv + vdp)$ for $d\theta$ in (21): hence
$$\begin{aligned}dQ &= \frac{C_v(pdv + vdp)}{R} + pdv \\ &= \frac{C_v + R}{R} pdv + \frac{C_v}{R} vdp \\ &= \frac{C_p}{R} pdv + \frac{C_v}{R} vdp \end{aligned} \quad \ldots\ldots\ldots(23).$$

44. The gas will generally be losing or gaining heat by conduction through the sides of the cylinder, but if a good non-conducting material be used either in making the cylinder or as a covering for it, the rate at which heat is lost or gained in this way will become very small. We are thus led to imagine an ideal case in which no heat at all can enter or leave the gas during the expansion. We have then $dQ = 0$, and therefore, by equation (21),

$$C_v \frac{d\theta}{\theta} + R \frac{dv}{v} = 0.$$

Integrating,

$$C_v \log \theta + R \log v = \text{a constant.}$$

Writing $C_p - C_v$ for R, and k for $\frac{C_p}{C_v}$, we get

$$\log \theta + (k - 1) \log v = \text{a constant,}$$

or $\theta v^{k-1} = \text{a constant.}$

Hence if θ_0, v_0 are the values of θ, v at any instant, we shall have

$$\frac{\theta}{\theta_0} = \left(\frac{v_0}{v}\right)^{k-1} \quad \ldots\ldots\ldots\ldots\ldots\ldots(24).$$

Suppose, for example, that a quantity of air, originally at the freezing point, is contained in a cylinder impermeable to heat, and let the piston be slowly pushed in till the volume is reduced by half: then $\theta_0 = 273$ and $k = 1\cdot410$, whence

$$\frac{\theta}{273} = 2^{\cdot 410} = 1\cdot329,$$

and $\theta = 273 \times 1\cdot329 = 363$, or $\theta' = 363 - 273 = 90$.

The compression therefore causes the temperature to rise 90° C.

Again, if $v = \frac{v_0}{4}$, and $\theta_0 = 273$, as before, we have

$$\frac{\theta}{273} = 4^{\cdot 410} = 1\cdot 765,$$

and $\theta = 273 \times 1\cdot 765 = 482$, or $\theta' = 482 - 273 = 209$.

To find the relation between θ and p, we may either substitute for v in equation (24) by means of the relation $pv = R\theta$: thus

$$\frac{\theta}{\theta_0} = \left(\frac{p}{p_0}\frac{\theta_0}{\theta}\right)^{k-1},$$

so that
$$\left(\frac{\theta}{\theta_0}\right)^k = \left(\frac{p}{p_0}\right)^{k-1} \quad \ldots\ldots\ldots\ldots\ldots\ldots(25);$$

or we may work from equation (22): thus

$$C_p d\theta - R\frac{\theta}{p} dp = 0,$$

whence $\quad C_p \log \theta - (C_p - C_v) \log p = \text{a constant},$

or $\quad k \log \theta - (k - 1) \log p = \text{a constant},$

or $\quad \dfrac{\theta^k}{p^{k-1}} = \text{a constant}.$

Hence, just as before,

$$\frac{\theta^k}{p^{k-1}} = \frac{\theta_0^k}{p_0^{k-1}}.$$

The relation between p and v is

$$C_v v dp + C_p p dv = 0,$$

that is,
$$C_v \frac{dp}{p} + C_p \frac{dv}{v} = 0.$$

Hence $\quad \log p + k \log v = \text{constant}.$

If p_0, v_0 are corresponding values of p and v, this becomes

$$pv^k = p_0 v_0^k,$$

ON PERFECT GASES.

or
$$\frac{p}{p_0} = \left(\frac{v_0}{v}\right)^k \qquad \ldots\ldots\ldots\ldots(26).$$

The work done by the gas in altering its volume from v_1 to v_2 is

$$\int_{v_1}^{v_2} p\,dv = p_0 v_0^k \int_{v_1}^{v_2} \frac{dv}{v^k}$$

$$= \frac{p_0 v_0^k}{k-1} \left\{ \frac{1}{v_1^{k-1}} - \frac{1}{v_2^{k-1}} \right\}.$$

When any substance is slowly compressed or rarefied in a cylinder impermeable to heat, the corresponding indicator diagram is called by Rankine an Adiabatic curve, and by Prof. Gibbs, an Isentropic curve, because the Entropy, a quantity which will be discussed in the next chapter, remains constant throughout the operation. On the other hand, when the temperature is kept constant during the process, the corresponding diagram is known as an Isothermal curve.

The equation of an adiabatic curve for a perfect gas being

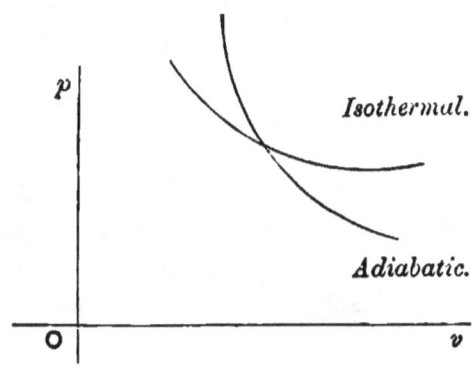

$$pv^k = \text{constant},$$

and that of an isothermal

$$pv = \text{constant},$$

100   ELEMENTARY THERMODYNAMICS.

it is clear, since k is > 1, that if an isothermal and an adiabatic curve cross one another at any point, the adiabatic curve will be more steeply inclined to the axis of v at the point of intersection than the isothermal.

45. We are now in a position to consider a very important example of the conversion of heat into mechanical work, which will prepare the way for the next chapter.

Let us suppose a gramme of any perfect gas contained in a cylinder fitted with an air-tight piston, as shown in the sketch, the piston and every part of the cylinder except the bottom being impermeable to heat, while the bottom is supposed to have so small a capacity for heat

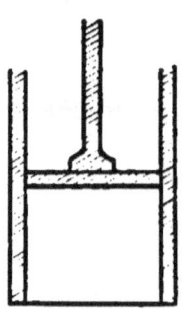

that the heat required to raise its temperature may be neglected. Such a cylinder cannot be constructed in practice. It is merely a limit towards which we may approximate very closely, but which can never be actually attained.

Let us also suppose that there are two bodies A, B of any kind, the temperatures of which, as shown on the air thermometer, are kept constantly equal to θ_a, and θ_b,

ON PERFECT GASES. 101

respectively; B being colder than A, and of the same temperature as the gas inside the cylinder. Also let C be

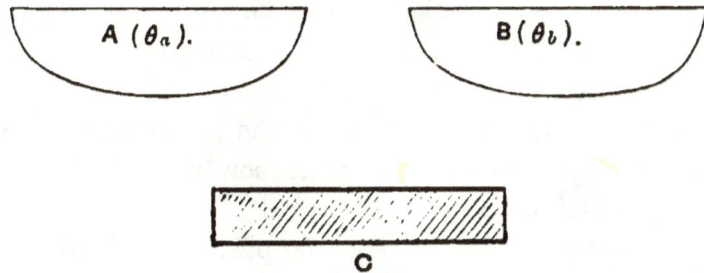

a perfect non-conductor on which the cylinder can be set for any length of time without losing or gaining heat.

Then let the gas inside the cylinder be made to undergo the following complete cycle of operations, the conception of which is due to Carnot.

(1) Let the cylinder be placed on C, and then press

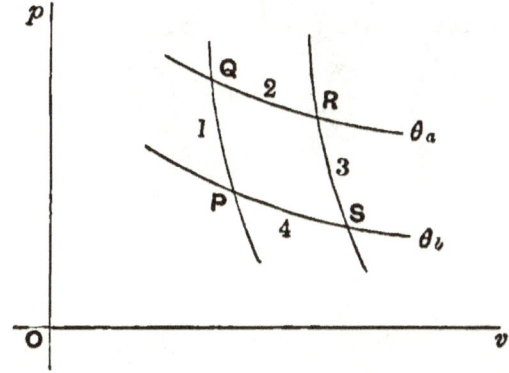

the piston down very slowly until the temperature of the gas rises from θ_b to θ_a, the operation being represented on the indicator diagram by the adiabatic curve PQ.

(2) Transfer the cylinder to A and then let the piston

be drawn out so slowly that the temperature of the gas is prevented from falling by the passage of heat from A through the bottom of the cylinder. Let the heat absorbed by the gas be called q_a, and suppose the operation represented on the diagram by the isothermal curve QR.

(3) Place the cylinder again on C and then slowly draw the piston further out until the temperature of the gas again becomes θ_b. The corresponding curve in the figure is the adiabatic RS.

(4) Lastly, let the cylinder be placed on B, and then by forcing in the piston slowly let the volume be diminished at the constant temperature θ_b until the initial state of the gas is attained, the heat given to B being denoted by q_b.

The heat, q_a, which is absorbed by the gas in the second operation is the thermal equivalent of the work done by the gas in that operation, since the temperature is constant. Hence if v_q, v_R be the volumes of the gas corresponding to the points Q, R, we have

$$q_a = \int_q^R p\, dv = R\theta_a \int_q^R \frac{dv}{v}$$

$$= R\theta_a \log \frac{v_R}{v_q}.$$

Similarly $$q_b = R\theta_b \log \frac{v_s}{v_p}.$$

Now since the equation to an adiabatic curve is

$$\theta v^{k-1} = \text{a constant},$$

we get, from the adiabatic curve PQ,

$$\frac{\theta_a}{\theta_b} = \left(\frac{v_p}{v_q}\right)^{k-1},$$

and from the adiabatic curve RS

$$\frac{\theta_a}{\theta_b} = \left(\frac{v_s}{v_R}\right)^{k-1}.$$

Hence
$$\frac{v_P}{v_Q} = \frac{v_S}{v_R},$$

or
$$\frac{v_R}{v_Q} = \frac{v_S}{v_P},$$

and therefore
$$\frac{q_a}{\theta_a} = \frac{q_b}{\theta_b} \quad \ldots\ldots\ldots\ldots\ldots\ldots(27).$$

The total work done by the gas during the cycle will be $q_a - q_b$, by the first law of thermodynamics, and since the process may be repeated indefinitely, we may consider that this work has been transformed out of heat.

The ratio of the work obtained from the cycle to the mechanical equivalent of the heat absorbed from the hotter body, is $\frac{q_a - q_b}{q_a}$, or $1 - \frac{\theta_b}{\theta_a}$, and is therefore the same whatever kind of perfect gas be operated upon. It is called the 'efficiency' of the cycle.

CHAPTER III.

CARNOT'S PRINCIPLE.

46. ACCORDING to the principle of the Conservation of Energy, the total energy of any material system which is prevented from exchanging energy with external systems and from radiating energy into infinite space, remains invariable. But though the total energy continues the same in amount, it may assume different forms, energy being convertible, under the proper circumstances, out of any one form into an equivalent in any other form.

The foundations of the important and interesting subject of the Transformation of Energy were laid by Carnot in a profound essay published in 1824. This remarkable and curious work was, unfortunately, vitiated by the false view then prevalent as to the nature of heat. It is satisfactory, however, to find that before he died in 1832, Carnot had not merely emancipated himself from the doctrine of caloric[1], but had made a good approximation to the mechanical value of a calorie[2].

Although Carnot is now admitted to have been one of

---

[1] See Bertrand's Thermodynamics.

[2] Carnot concluded that a calorie was equivalent to 37000 gramme-centimetres. This is a more accurate estimate than Mayers (36500) obtained in 1842.

See Carnot's 'Réflexions sur la puissance motrice du feu.' (Gauthier-Villars, Paris, 1878.)

the greatest men produced by France, and his principle to be one of the greatest discoveries yet made in science, his work failed to attract attention for some time after it was published. But ten years later, it was brought prominently forward by Clapeyron who exhibited it in a more elegant form by means of Watt's indicator diagram. Still very little progress appears to have been made with the subject until the appearance of the present theory of heat.

In 1850, soon after the establishment of the mechanical theory of heat, it was shown by Clausius that Carnot's reasoning could easily be modified so as to be consistent with the new theory and he then proceeded to recast it in its present form. In this undertaking he was ably seconded about the same time by Rankine, and a little later by Sir W. Thomson, to whom we owe a truly 'absolute' scale of temperature.

47. Before we can make use of Carnot's principle, it is necessary to introduce the conception of reversible operations. A few illustrations will make this idea clear.

(1) When two pieces of metal C, D are in contact at the same temperature without being rubbed together, no heat will pass from one to the other. But if the temperature of C be raised ever so little, heat will at once begin to flow across the junction, and if the small difference of temperature be maintained by any means for a sufficient time, any amount of heat may be transferred from one body to the other. In like manner, by slightly raising the temperature of D without altering that of C, a finite quantity of heat may be made to pass in the opposite direction. The reversible limit towards which we may approximate as near as we please without ever being able

actually to reach, is that of causing heat to flow in either direction without any difference of temperature at all.

(2) Let us suppose that a cylindrical rod, which is placed parallel to the axis of x, is made to move with an acceleration to the right. Then if P, Q be any two

sections of the rod, it is clear that the forces exerted on the portion PQ across the sections P and Q will not generally be equal. For example, if these forces be tensions and no other forces act parallel to the axis of x, the tension across the section Q must be greater than the tension across the section P. If, on the contrary, the rod moves with an acceleration to the left, the tension across the section Q will be less than that across P. The internal state of the rod is therefore different in these two opposite motions; but we may make the difference as small as we please without actually causing the accelerations to become zero. Then by allowing sufficient time, any finite change of velocity may be produced. The limit in this case which cannot be exactly attained, is that of causing the velocity parallel to the axis of x, of the portion PQ, to alter without any difference in the tensions across the sections P, Q. If, in this limiting case, there be no electrification and heat be imparted to or abstracted from the rod in a reversible manner, the operation which the rod undergoes will be completely reversible.

(3) If a quantity of gas be contained in a cylinder fitted with an air-tight piston, it is evident that the piston

cannot be pushed in without exerting a greater force on the gas than when it is at rest: similarly if the expansion of the gas force the piston out a little, the pressure of the gas on the piston will be less than if the piston had not moved. But in both cases, by making the velocity of the piston slow enough, we may approximate to an ideal operation in which the pressure of the gas on the piston is the same as if it was at rest. If, in this ideal case, the temperature of the gas be kept constant, or caused to depend only on the volume, it is clear that the gas may be compressed back again by exactly the same force which it has overcome in expanding.

(4) A remarkable example of reversibility may occur when a cylinder of moderate size is filled with water and steam instead of gas. If, as before, the piston be pushed in, equilibrium will be impossible within the cylinder, both as regards pressure and temperature, so long as the motion lasts. But if, after the motion of the piston ceases, the temperature be reduced to the same value as before and there be still room for steam to be formed, it is found by experiment that the pressure is unaltered by the compression. We are therefore led to imagine a limiting reversible case in which steam is converted into water or water into steam without any change either in the pressure or the temperature. The indicator diagram representing such a process will be a straight line parallel to the axis of v.

(5) Again, let us suppose the cylinder to contain a saturated aqueous solution of a salt which deposits the anhydrous salt and not a hydrate; for example, Nitrate of Lead. Also let there be a quantity of the salt undissolved. Then if the temperature be slowly changed, the solution

will alter in strength by diffusion and it is clear that the operation will ultimately be reversible.

In every case of reversibility, the process which the system undergoes is a non-frictional process; but it does not follow, conversely, that every non-frictional process is reversible. We have already had examples of irreversible non-frictional processes in Article 25, Chap. I. Again, if a cylinder contain a saturated solution of Sulphate of Magnesium together with an excess of the salt, increase of temperature will increase the strength of the solution by causing some of the salt to be dissolved; but on again reducing the temperature, the solution will deposit, not the anhydrous salt, but a hydrate. The process is therefore not reversible, even though it is a non-frictional process.

48. Carnot's principle relates to the transformation of heat into mechanical work.

Suppose that any material system whatever, which may be electrified in any way we please, is made to undergo a change of state during which there are no external electric influences, so that the energy cannot be affected by external systems except by the conduction or radiation of heat and by the doing of mechanical work. Then, unless the change of energy is zero, we cannot assert that the mechanical work obtained from the system has been transformed out of the heat absorbed by it, for part of the work may owe its origin to the change in the energy of the system. It will therefore be necessary for us, in the first instance, to restrict ourselves to the consideration of complete cyclical processes.

Again, if the system be gaining or losing heat at any finite speed in any part A, either by conduction or radiation, it will be impossible to keep the temperature of A strictly uniform and constant; but by taking sufficient precautions, we shall sometimes be able to prevent the temperature of every part of A from differing by more than an infinitesimal amount from a given constant temperature. For example, if A be a thin closed metallic vessel filled with a very volatile liquid and its vapour, any difference in the temperatures of different parts of A would cause an explosive formation or condensation of vapour, by which the temperature of A would quickly be rendered uniform. Also if A consist of two parts of which the upper part is a cylinder fitted with an air-tight piston and connected by a pipe with the lower part, it will be practically possible, by moving the piston in or out, to keep the temperature of A not merely uniform, but constant.

Let us now suppose the system to undergo any cyclical process during which it is prevented from receiving or losing heat, either by conduction or radiation, except in two parts A, B. Then as we approach the ideal case in which the temperatures of A and B are kept uniform and constant, we assume as an axiom that, whatever the two parts A, B may be, no mechanical work can be obtained from the system during the cycle so long as the temperatures of A and B are equal. If work is gained from the cycle, the temperatures of A and B must therefore be different.

This important fundamental axiom is substantially due to Carnot and its great merit arises from the fact that the operations referred to constitute a complete cycle. It is not generally true when the cycle is incomplete, as will appear later on. The preceding is probably the simplest way in which it can be stated; but in this book we shall follow the usage of all existing text-books by employing it in a form differing but little from that in which it was originally stated by Carnot himself. We shall suppose that the system undergoes a complete cycle such that all the heat absorbed, whether by conduction or radiation, comes from two bodies A, B, whose temperatures are kept uniform and constant, and that conversely, all the heat given out goes to the same two bodies A, B; and we shall assume as an axiom that no mechanical work can be obtained from the system during the cycle so long as the temperatures of A and B are equal.

If the given system can lose or gain heat by conduction only, it will be clearly possible to keep the temperatures of A and B practically uniform and constant by causing all the heat exchanged between the system and A and B to be conducted respectively through two immense intermediate bodies, or systems of bodies, A', B'. The condition that the whole of the heat lost from the given system should be transmitted by A' and B' to A and B, and conversely, can easily be satisfied. We take A' and B' unelectrified, and consequently free from all external electric influences; and then make them both undergo complete cyclical processes during the same time as the given system, without either doing work upon them, or obtaining work from them. In these cycles, the total amount of heat absorbed by either A' or B' is obviously zero. Hence if we suppose that both A' and B' are prevented from radiating energy themselves, and from receiving the radiation of other bodies; and also that A' can only exchange heat by conduction with the given system and with A, B' with the given system and with B;

it follows from Art. 15, that if we neither rub the given system nor A nor B against A' or B', the quantities of heat conducted during the cycle from the given system into A' and B' will be respectively equal to the quantities of heat obtained by A from A', and by B from B', in the same time; and conversely.

If the given system can lose or gain heat by radiation only, we suppose that all the heat is exchanged by radiation, not with A and B directly, but with two intermediate bodies A', B', which are the same as before except that they now exchange heat with the given system by radiation. Also if we suppose that no radiation is transmitted across the spaces A'', B'', which separate the given system from A' and B', except the radiation passing between the system and A' and B', it is clear that if the spaces A'', B'' be in exactly the same condition at the end of the cycle as at the beginning of it, the quantities of heat lost by the given system to A' and B' during the cycle will be respectively equal to the quantities derived in the same time by A from A' and by B from B'; and conversely. Unless this condition is fulfilled, the system cannot be said to exchange heat with the two bodies A, B, only. For example, if the system and A', B' be at great distances from one another, and be prevented from emitting radiation until the beginning of the cycles which they undergo, these cycles may be completed before the radiation emitted by any one of them can get to the other: in this case, the system would be considered to undergo a cycle in which the whole of the heat given out was supposed to be radiated into infinite space.

Since length of time is the essential element in all non-frictional processes, with the sole exception of the non-frictional process by which the motion of the centre of mass is varied, which, however, cannot form a complete cycle unless the process is indefinitely slow, it is evident that if the cycle undergone by the given system be reversible, the cycles undergone by A' and B' will also be reversible; and consequently the system will then absorb or give out heat at the temperatures of A and B only.

49. The quantities of heat absorbed by the material system from the two bodies A, B, during any complete cycle cannot both be positive. For we could then, by expending work in friction, cause the system to undergo a cycle of operations in which a positive quantity of heat was absorbed from one of the bodies A, B, and no heat at all either received from or parted with to the other. In other words, we should be able to take heat from a body whose temperature was uniform and constant, and transform it into work without the presence of any other body of different temperature, contrary to Carnot's axiom.

If the cycle be reversible, the quantities of heat absorbed from A and B cannot both be negative; for by simply reversing the cycle we should obtain another cycle in which the quantities of heat absorbed were both positive. Hence, in any reversible cycle, a positive quantity of heat is taken from one of the bodies A, B, and a positive quantity given to the other. Consequently, if q_a be the heat absorbed from A in a reversible cycle and q_b the heat given to B, q_a and q_b will be of the same sign and their ratio positive. If the cycle be irreversible, q_a and q_b may either be of the same or of opposite signs and their ratio either positive or negative.

Again, suppose that any two material systems whatever, X, X', undergo any complete cycles of operations during which they can only exchange heat with the two bodies A, B, and let q_a, q_a' be respectively the quantities of heat absorbed from A, and q_b, q_b' the quantities restored to B. Then if the cycle undergone by X be reversible, the ratio $\dfrac{q_a'}{q_b}$ cannot be greater than the positive ratio $\dfrac{q_a}{q_b}$ when q_b' is positive, nor less than $\dfrac{q_a}{q_b}$ when q_b' is negative. For if the cycle undergone by X be reversed, the system X will then absorb q_b from B and give back q_a to A. We may therefore by increasing or decreasing the system in the proper proportion, cause it to absorb q_b' from B and give up $q_a \dfrac{q_b'}{q_b}$ to A. Combining this new cycle with that of X', we get a new cycle Y, in which no heat is exchanged with B while a quantity $q_a' - q_a \dfrac{q_b'}{q_b}$ is absorbed from A. Hence, by Carnot's axiom, $q_a' - q_a \dfrac{q_b'}{q_b}$ cannot be positive, and therefore $\dfrac{q_a'}{q_b}$ cannot be greater than $\dfrac{q_a}{q_b}$ when q_b' is positive, nor less when q_b' is negative.

If the cycle undergone by X' be also reversible, the cycle Y will be reversible, and, by simply reversing it, we may show that $q_a' - q_a \dfrac{q_b'}{q_b}$ cannot be negative. Hence $q_a' - q_a \dfrac{q_b'}{q_b} = 0$, so that, when both cycles are reversible,

$$\frac{q_a'}{q_b'} = \frac{q_a}{q_b} \quad \ldots\ldots\ldots\ldots\ldots\ldots (28),$$

whatever be the natures of the systems X, X', or of the reversible cycles which they undergo.

If the system X undergo a reversible cyclical process during which it can only exchange heat with two other bodies C, D, whose temperatures are kept uniform and constant, it is easily proved that, whatever the bodies C, D may be, if the temperature of C be equal to that of A and the temperature of D equal to that of B, we shall have

$$\frac{q_c}{q_d} = \frac{q_a}{q_b},$$

where q_c is the heat absorbed from C and q_d the heat given to D.

The ratio $\frac{q_a}{q_b}$ for a reversible cycle can therefore depend only on the temperatures of A and B, and is independent of:—

(1) The natures of the bodies A and B.

(2) The nature of the system which undergoes the reversible operation.

(3) The nature of the reversible operation itself.

50. The different temperatures of bodies may be conveniently distinguished by applying a different number to each temperature. These numbers are chosen by Sir W. Thomson in such a way that the numbers θ_a, θ_b, corresponding to any two temperatures A, B, satisfy the relation

$$\frac{q_a}{\theta_a} = \frac{q_b}{\theta_b} \quad \ldots\ldots\ldots\ldots\ldots (29),$$

where q_a is the heat absorbed from A and q_b the heat restored to B in any complete reversible cycle working between A and B. We are then at liberty to fix the value of any one temperature we please, or, better still, we may choose the numbers so that the temperatures of the ordinary freezing and boiling points differ by 100. Such a scale of temperature is known as Thomson's absolute scale, because it is independent of the particular properties of any substances or class of substances. It obviously coincides very nearly with the scale of the air thermometer, or with that of the centigrade mercury thermometer increased by 273. The freezing point on Thomson's absolute scale will therefore be about 273 and the boiling point 373. Also since q_a and q_b are of the same sign, it is evident that all the numbers which denote temperatures on Thomson's scale are positive.

It should be noticed that the numbers which distinguish temperatures might have been chosen in many different ways so as to be consistent with Carnot's principle, but we could not have chosen them arbitrarily. Thus we could not have taken θ_a and θ_b so that

$$\frac{q_a}{q_b} = \frac{\theta_a}{1+\theta_b};$$

for then we should also have had $\frac{q_b}{q_c} = \frac{\theta_b}{1+\theta_c}$ and $\frac{q_a}{q_c} = \frac{\theta_a}{1+\theta_c};$ which would have involved a contradiction.

On the ordinary thermometers, the different temperatures are distinguished by different numbers, the choice of which depends on some special property of the thermometric substance which is not possessed exactly by any other substance. Hence if two thermometers be constructed with different thermometric substances and be graduated in the usual way, so that they show the same readings at the freezing and boiling points, they will not generally agree at other temperatures.

For the future, when we speak of temperature, we shall

always understand Thomson's absolute scale, unless it is specially stated otherwise.

If we put $\theta_b = 0$, equation (29) becomes

$$\frac{q_a}{\theta_a} = \frac{q_b}{0},$$

from which we deduce $q_b = 0$. This result probably means that at the absolute zero of temperature, the system possesses no non-mechanical kinetic energy. For example, ice is nearly incompressible and its specific heat may be supposed independent of the pressure and temperature and equal to ·5 calorie. Hence, since no change of aggregation occurs between absolute zero and the melting point, the non-mechanical kinetic energy[1] of a gramme of ice at 0° C. may be taken to be about 136·5 calories, or 5,670 million ergs, or 415 foot-pounds. In English measure, the non-mechanical kinetic energy of a pound of ice at 0° C. will be about 190,000 foot-pounds. But if every particle be moving with the same non-mechanical velocity of v centimetres per second, the non-mechanical kinetic energy of a gramme will be $\frac{1}{2}v^2$ ergs: hence

$$v^2 = 11,340 \times 10^6,$$

or
$$v = 106,500.$$

Thus the non-mechanical velocity of each particle is 106,500 centimetres, or 1·065 kilometres, or 1164 yards, per second. This is equal to a velocity of 63·9 kilometres, or nearly 40 miles, per minute. In the case of iron, the

---

[1] When a solid body is raised in temperature without being melted, the increase of energy is probably, to a large extent, non-mechanical kinetic energy, and, in a rough calculation like the present, may be taken to be entirely such.

specific heat is about $\frac{1}{6}$th of a calorie, and therefore the non-mechanical kinetic energy of a gramme of iron at 0° C. is about 1,260 million ergs. Thus the non-mechanical velocity of each particle is about 50,000 centimetres, or ·5 kilometres, or 547 yards, per second; that is, 30 kilometres, or nearly 19 miles, per minute.

The non-mechanical motions are too minute to be observed, even with the most powerful microscope, but it appears from theory that the maximum non-mechanical displacement of a particle in a solid is probably less than the 200 millionth of a centimetre (500 millionth of an inch). Hence since each particle of a solid describes an oval of some sort, the non-mechanical motions in iron at 0° C. must be reversed in any given direction at least 10^{12} times in a second.

51. In a complete cycle working between A and B, the quantities q_a and q_b may be either positive or negative; but since a positive quantity of heat cannot be taken both from A and B, it is evident that heat can only be transformed into work during the cycle when a positive quantity of heat is taken from one of the two bodies A, B, and a positive quantity given to the other, that is, when q_a and q_b are of the same sign. In discussing the conversion of heat into work in this article, we shall not impair the generality of our reasoning if we suppose that q_a and q_b are both positive; in consequence of which the ratio $\frac{q_a}{q_b}$ cannot be greater than $\frac{\theta_a}{\theta_b}$.

If θ_a be greater than θ_b, the body A is said to be at a higher temperature than B. Hence we see that, in a complete reversible cycle, a positive quantity of heat may be taken from any body A and partially converted into work provided that there is a second body B, of lower temperature, to absorb the waste heat. The hotter body A is known as the 'source' and the cooler body B as the 'refrigerator.'

The heat which is converted into work is $q_a - q_b$ and its ratio to the heat absorbed from the hotter body is $\dfrac{q_a - q_b}{q_a}$, which, since the cycle is reversible, is equal to $1 - \dfrac{\theta_b}{\theta_a}$. This ratio, which is called the 'efficiency' of the cycle, depends therefore only on the temperatures of the source and refrigerator.

If the cycle be not reversible, the ratio $\dfrac{q_a}{q_b}$ will either be less than, or at most, equal to $\dfrac{\theta_a}{\theta_b}$. It will be easily seen that if the irreversible cycle be non-frictional, $\dfrac{q_a}{q_b}$ will be equal to $\dfrac{\theta_a}{\theta_b}$, and that in all other cases it will be less. Thus, when a cycle is frictional, its efficiency, which is the ratio $\dfrac{q_a - q_b}{q_a}$, or $1 - \dfrac{q_b}{q_a}$, will be less than $1 - \dfrac{\theta_b}{\theta_a}$, the efficiency of a non-frictional cycle working between the same source and the same refrigerator. In popular language, no cycle can transform so large a proportion of heat into work as a non-frictional[1] cycle.

If θ_a be less than θ_b, q_a will also be less than q_b. In this case, in the course of a complete reversible cycle, a

---

[1] Here, and in much that follows, the term 'friction' is used in a general way to include every kind of irreversibility as well as friction proper.

positive quantity of heat is taken from the cooler body and a *larger* quantity given up to the hotter. Thus an expenditure of work is necessary during a reversible cycle if we wish to lift a quantity of heat from a lower to a higher temperature, and the amount of work required is the same for all material systems and for all reversible cycles. When the cycle is not reversible, $\dfrac{q_a}{q_b}$ cannot be greater than $\dfrac{\theta_a}{\theta_b}$. Hence if q_a have the same value as in a reversible cycle, q_b will have a greater, or, in the case of non-frictional irreversible cycles, an equal value. Consequently, heat cannot be raised from a lower to a higher temperature with so small an expenditure of work as when the cycle is non-frictional.

It is not the least important consequence of Thomson's system of temperature that heat cannot flow of itself, either by conduction or radiation, from a lower to a higher temperature. But when a thermometer, graduated in the usual way, is employed to show differences of temperature, the thermometric substance may be such that heat sometimes appears to pass from a colder to a hotter body. For example, if within certain limits of temperature the thermometric substance contracts as its temperature, as shown by the air thermometer, increases, like water between 0° C. and 4° C., and if A, B be any two bodies whose temperatures lie within these limits; then if the temperature of A, as shown by the air thermometer, be higher than that of B, heat will be able to pass of itself from A to B and yet the temperature of B would be shown by the thermometer we have just described to be higher than that of A.

52. It has hitherto been usual to consider only the case in which the material system consists of a single unelectrified body without mechanical motions whose temperature when in a state of equilibrium is the same throughout. Such a body when it undergoes a reversible cycle of any kind is known as a 'Carnot's perfectly reversible engine.'

In Carnot's time, heat was believed to be a material substance which could neither be created nor destroyed, and the total quantity of heat contained in any system was supposed to depend only on its state. It thence followed that in any cyclical process, exactly as much heat was absorbed as given out, and it was thought that the work obtained from the cycle was done at the expense of the fall of temperature; just as the work done by a stream in driving a water-wheel depends on a change in the level of the stream. Yet notwithstanding these false assumptions, it was easily shown, as in Article (49), that if W be the work obtained from the engine during a complete cycle, and Q the heat absorbed from the source or given to the refrigerator, the ratio $\dfrac{W}{Q}$ was the same for all reversible engines working between any sources and refrigerators whose temperatures were respectively equal to those of two given bodies A, B. It could therefore depend only on the temperatures of A and B. In addition to this conclusion, Carnot also obtained one of the most important results of the next chapter.

53. Carnot's principle includes two of the three experimental results given in elementary books as The laws of Friction. The third cannot be deduced from Carnot's principle, because it is not rigidly accurate, although generally a very good approximation.

Suppose that we have two bodies A, B, of the same temperature, which possess no potential energy, or at least, only a constant amount, and let them be rubbed together by the hand in such a way that the other external forces do no work. Then if the mechanical motions be zero both at the beginning and at the end of the operation, the cycle may be completed by allowing the two bodies to exchange heat with some third body C. Hence, by Carnot's axiom, there is an expenditure of mechanical work during the cycle, and therefore we conclude that:—

I. Friction acts on each body in a direction opposite to that in which its relative motion takes place, or merely tends to take place.

II. No more friction can ever be called into play than is just sufficient to prevent relative motion.

54. It is sometimes, but erroneously, supposed that Carnot's principle asserts that it is impossible to raise the temperature of a body without an expenditure of work. The following two examples will illustrate this and similar points.

(1) If the bulb of a thermometer be wrapped round with flannel and then breathed on, the temperature of the thermometer is found to rise higher than the temperature of the breath. This experimental fact will occasion no difficulty if we remember that Carnot's axiom asserts nothing unless the changes of state constitute a complete cycle, and that even then, the only temperatures taken into account are the temperatures of the two

bodies from which heat is received or to which heat is lost.

(2) The following ingenious operation was described by Hirn in 1862. Let there be two similar and equal cylinders surrounded by good non-conducting materials and connected at the bottom by a very narrow copper pipe. Also let the cylinders be fitted with air-tight pistons the rods of which are constructed with equal teeth to engage a spur wheel so that when one piston rises the other will descend an equal distance. The whole space beneath the pistons will then be invariable, because, as the space under one piston decreases, the space under the other increases by an equal amount.

Now let us suppose that initially the piston B is at the bottom of its cylinder and the other cylinder filled

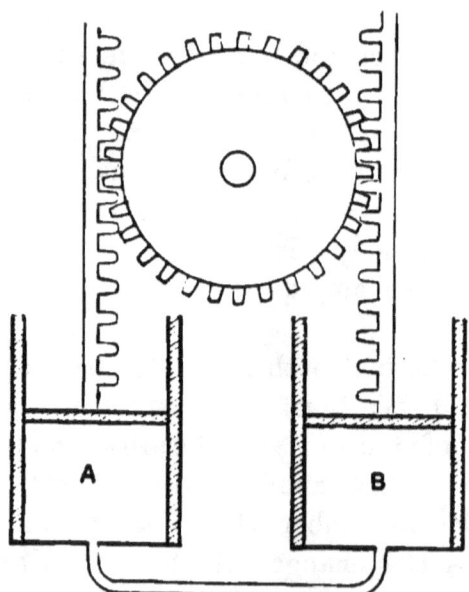

with gas at $0°$ C., the pipe being kept surrounded by the

steam of boiling water. Then let the wheel be slowly turned so that the piston A is caused to descend and B to rise. If the pistons be well made, very little mechanical work will be required but important effects will be produced on the gas. For as soon as the motion commences, a small quantity of gas will pass through the pipe where its temperature will be raised from $0°$ C. to $100°$ C. In consequence its volume will increase and the gas which remains in A will be somewhat compressed and its temperature slightly raised. As the next small quantity of gas passes through the pipe, it will be heated and will expand in the same way, and thereby the gas contained in both cylinders will be still further compressed and raised in temperature. The motion of the wheel being continued until the piston A gets to the bottom of the cylinder, the whole of the gas will have passed through the pipe into the other cylinder and the temperature of every part will be $100°$ C. or higher, the mean temperature being about $120°$ C. Hence in completing the cycle by bringing the temperature of every part of the gas back to $0°$ C., heat may pass out of the gas to bodies whose temperatures are considerably above $100°$ C. We are therefore able to raise heat from a lower to a higher temperature without an expenditure of work. But it must be remembered that Carnot's principle supposes that there are only *two* external bodies with which the system can exchange heat. In the present instance, the cycle cannot be completed unless there are bodies of as low a temperature as $0°$ C. to which the gas may part with its heat. Examples like this cannot be fully discussed until we have considered the case in which there are any number of external bodies with which a given system can exchange heat.

55. When any material system is subjected to external electric influences, it is easily proved in works on electricity that the system may gain or lose 'electric energy' (electric work) without either gaining or losing heat. If the system undergo a cyclical process during which there is neither absorption nor evolution of heat, the whole of the work obtained during the cycle will have been transformed out of electric energy, so that electric energy may be transformed wholly into mechanical work, and, conversely, mechanical work may be transformed wholly into electric energy. Hence, by Carnot's principle, electric energy may be transformed wholly into heat, but heat can only be transformed partially into electric energy.

Any system which absorbs or loses electric energy may be called an electric, or an electro-magnetic, engine, and it is evident that it is not subject to the restrictions of Carnot's principle. For the future, we shall suppose that there are no external electric influences, so that Carnot's principle will always be true.

56. When any material system is enclosed in a vessel of any kind which is protected from all external influences, mechanical, thermal, and electrical (including the radiation of energy into external space as an external influence), it may for the present be assumed as an extension of Carnot's principle, or of the principle of friction, that the system will ultimately attain a state in which all its visible properties are invariable and the temperatures of all its parts equal. The whole system will then behave as a rigid body, so that the only[1] mechanical motions it

[1] See Art. 24.

can have consist of a uniform motion in a straight line of the centre of mass combined with a constant angular rotation of the whole system about some straight line passing through the centre of mass in a constant direction. Also since radiation depends on the non-mechanical motions which always exist except when the absolute temperature is zero, it may be further assumed that if the given system had been at liberty to radiate energy into infinite space but had otherwise been protected as before, it would still have ultimately attained an invariable state, but the final absolute temperature of every part would have been zero.

It should be observed that if the system consist of several parts the temperatures of which are originally different, it is not asserted that the colder parts cannot become as hot as the hotter until the final state is reached. Suppose, for example, that the system consists of an unelectrified part A at $130°$ C. and an unelectrified part B at $70°$ C., both parts being without mechanical motions, and let B be a cylinder in which a quantity of gas is confined by an air-tight piston pressed upon by a strong spring but prevented from being forced in by a fusible plug which melts at $94°$ C. Then if A and B be placed in presence of one another in a vessel which is protected from all external influences and has so small a capacity for heat that its cooling or heating effects on A and B are very small, it is clear that A will cool and B get hotter until the melting of the fusible plug releases the spring. When this occurs, the temperature of B may rise to $200°$ C. or more, and then B would begin to cool and A would get hotter.

57. When a quantity of radiation falls on a body, it is easily shown by experiment that part of the radiation may be absorbed, part reflected, and the remainder transmitted; but since the tendency to equalisation of temperature is universal, it is evident that no body can reflect or transmit the whole of the radiation which falls upon them. There is nothing in Carnot's principle, however, to prevent us from thinking that a body may absorb the whole of the radiation which falls upon it, and in fact, any body well coated with lamp-black practically does so. We shall accordingly suppose that a surface of lamp-black absorbs the *whole* of the incident radiant energy. This assumption is made merely as a convenience and not to prove any properties which could not be proved without it.

Again, if a piece of ice be brought near the bulb of a thermometer, the temperature of the thermometer is observed to fall; but when a piece of red-hot iron is brought near, the temperature rises. These facts have led to Prevost's theory which has been fully developed by Stewart and others under the name of the Theory of Exchanges. It is supposed that every system is constantly radiating energy at a rate which depends only on its state and is independent of the presence or absence of other systems. In the two cases just given, heat is radiated from the thermometer faster than it is received from the ice and not so fast as it is received from the red-hot iron.

In applying Carnot's principle to radiation, we shall restrict ourselves to the case of unelectrified bodies without mechanical motions. The body B which is to be considered will be supposed suspended by a fine thread or string in a closed vessel V, of any shape or size, from which the air is exhausted, so that radiation can only pass between the body B and the containing vessel V across a vacuum; and it will be assumed, as usual, that radiant energy travels in a vacuum in straight lines in directions which can neither be deflected nor reversed except by some material obstacle. We shall also suppose that the vessel V is protected from all external

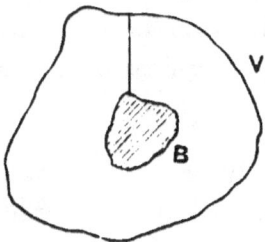

influences, so that V and B ultimately assume the same constant uniform temperature.

When B has attained its final state, it is evident that it must be absorbing exactly as much radiation as it emits. But if Q be the quantity of radiation which falls on B per second, one part, q_r, will be reflected, another part, q_t, transmitted, and the remainder, q_a, absorbed, where

$$q_r + q_t + q_a = Q,$$

so that the amount of absorption or emission per second is given by

$$q_a \equiv Q - q_r - q_t,$$

all the quantities being positive.

Now if B had been coated with lamp-black, q_r and q_t would both have been zero, and the emission per second would have been equal to Q, which is greater than in any other case. This result is usually expressed by saying that good reflectors or transmitters are bad radiators. It must be remembered, however, that this property is only shown to be true when B has attained its final state: thus if B reflect well at 50° C., it does not necessarily radiate badly at all other temperatures.

Again, since the radiation emitted by B per second when its surface is coated with lamp-black is equal to the whole radiation which falls upon it, it is clear that the incident radiation in this case is independent of the position of B inside V, and also of the nature and shape and size of V.

58. When a body is coated with lamp-black, the emission of radiation will be a surface phenomenon, because all the radiation coming from the interior will be absorbed by the lamp-black surface. Hence the emission per unit of area per second at any point of the surface will be independent of the curvature of the surface at that point. Consequently, if dq be the radiant energy emitted into a vacuum per unit of area per second at any point P of the surface within a small right cone of solid angle $d\omega$ whose axis makes an angle θ with the normal at P, the limiting value of $\dfrac{dq}{d\omega}$ will be a function of θ only. If therefore $d\sigma$ be a small area of the surface at P, the radiation sent out into a vacuum by this area within the small cone in a second will be

$$d\sigma \, d\omega \, f(\theta) \quad \ldots\ldots\ldots\ldots\ldots\ldots(30).$$

From this result we may obtain some very important conclusions.

(1) Let us take a closed vessel V whose interior is in the form of a hemisphere of radius l with a fine circular

cylindrical recess, the axis of which is at right angles to the base of the hemisphere and passes through the centre O, as in the figure. Let the recess be nearly filled by a plug B, one of whose ends is in the same plane as the base of the hemisphere. Also let both the interior of V and the whole of B be coated with lamp-black and then exhaust the air from V.

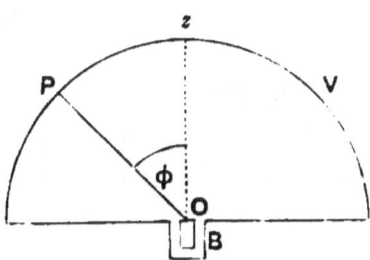

Then if any line OP be drawn from O to the surface of the hemisphere making an angle ϕ with the perpendicular Oz through O to the base, and if da be the area of the end of B, the solid angle it subtends at P will be

$$\frac{da \cos \phi}{l^2}.$$

Hence the radiation sent per second in the direction of the end of B by that part of the hemispherical surface included between the two right cones ϕ, $\phi + d\phi$, described about Oz, will be

$$2\pi l^2 \sin \phi d\phi \frac{da \cos \phi}{l^2} f(0),$$

or $\qquad 2\pi f(0) \, da \sin \phi \cos \phi d\phi.$

Of this radiation, the whole may reach B or part may be lost by friction in the ether. It is therefore necessary to denote the part which falls on B by

$$F\{l, \; 2\pi f(0) \, da \sin \phi \cos \phi d\phi\},$$

or, expanding in powers of da by Maclaurin's theorem,

$$2\pi f(0)\, da\, \psi(l) \sin\phi \cos\phi\, d\phi,$$

where it is obvious that $\psi(l)$ can have no constant value but unity.

Thus the total radiation which falls on the end of B per second is

$$2\pi f(0)\, da\, \psi(l) \int_0^{\frac{\pi}{2}} \sin\phi \cos\phi\, d\phi,$$

or
$$\pi f(0)\, da\, \psi(l).$$

This is equal to the emission from the end of B per second, and is therefore independent of l: thus $\psi(l) = a$ constant and *the ether is frictionless*.

(2) Again, suppose that V and B are concentric blackened spheres of radii R and r, respectively, and let the air be exhausted from V, as before. Then the radiation sent to B per second by each unit of area of V is

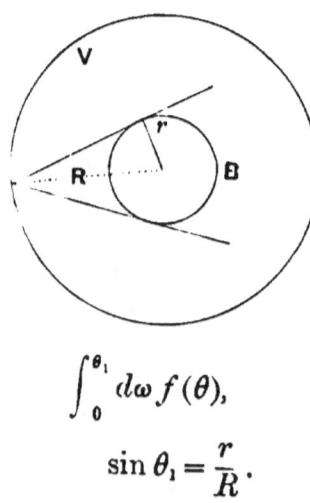

$$\int_0^{\theta_1} d\omega\, f(\theta),$$

where
$$\sin\theta_1 = \frac{r}{R}.$$

But the radiation emitted by B is proportional to its area, and equal to $4\pi\lambda r^2$, say. Hence

$$4\pi R^2 \int_0^{\theta_1} f(\theta)\, d\omega = 4\pi\lambda r^2,$$

or

$$2\pi \int_0^{\theta_1} f(\theta) \sin\theta\, d\theta = \lambda \sin^2\theta_1.$$

Differentiating,

$$2\pi f(\theta_1) \sin\theta_1 = 2\lambda \sin\theta_1 \cos\theta_1,$$

so that

$$f(\theta) = \mu \cos\theta,$$

where μ is a constant.

Thus the emission into a vacuum per second by a small area $d\sigma$ within the cone $d\omega$ takes the form

$$\mu \cos\theta\, d\sigma\, d\omega \quad \ldots\ldots\ldots\ldots\ldots\ldots (31).$$

Consequently the radiation sent out into a vacuum per unit of area per second at any point P of the blackened surface within a right cone described about the normal at P with any finite semi-vertical angle $\theta_1 \left(< \dfrac{\pi}{2}\right)$ is

$$2\pi\mu \int_0^{\theta_1} \sin\theta \cos\theta\, d\theta \equiv \pi\mu \sin^2\theta_1 \ \ldots\ldots (32),$$

and the total radiation at P is $\pi\mu$. Calling this k, the emission within the cone $d\omega$ becomes

$$\frac{k}{\pi} \cos\theta\, d\omega.$$

(3) Since radiant energy travels in a vacuum in straight lines, if two bodies A, B, be separated by a vacuum, it will only be possible for one ray to pass from a point P of one to a point Q of the other. But if a material body L be interposed between A and B, it may happen that several of the rays which proceed from

P have their directions so bent in passing through L that they are concentrated on a very small area surrounding

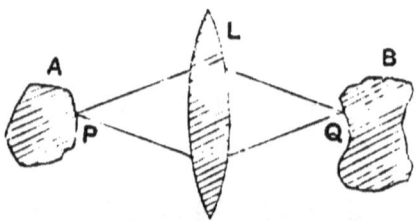

the point Q. One of the best examples of this is seen when the rays of the sun are brought to a focus by a lens (of glass or ice). If the lens is large enough, the effect produced may be considerable.

It will now be naturally asked whether it would not be possible, by means of a system of lenses, to cause heat to pass from a cooler to a hotter body without an expenditure of mechanical work, as was thought by Rankine? A careful examination of the question by Kirchhoff and Clausius has led them to the conclusion that such a result is impossible. To give a simple illustration, let us suppose two bodies A, B, and a lens L to be situated in a vacuum; and let us neglect the reflection and absorption that take place as the rays pass through L. Then if the focal

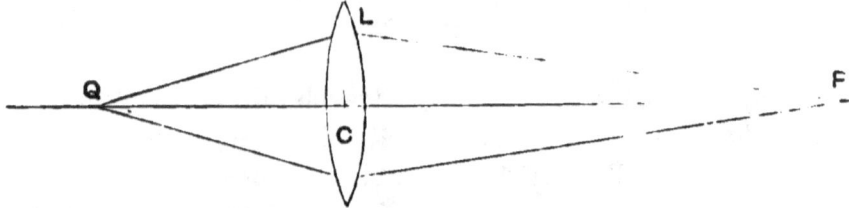

length of the lens be large, the rays which fall on it from a distant point P will converge with considerable accuracy to a *point* Q; and if PQ be the principal axis of the lens, the rays from any point p of a small area dS, placed at P at right angles to PQ, will, after passing through the lens, converge to a point q of a small area ds, placed at Q parallel to dS, the lines pq, PQ both passing through C, the optical centre of the lens.

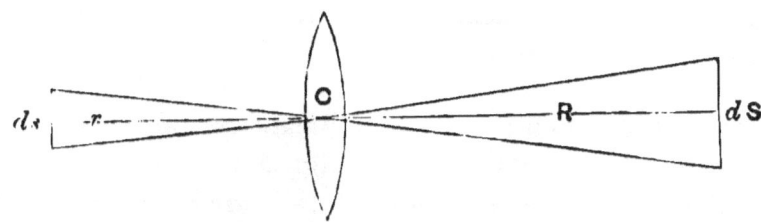

Now if R be the distance CP and r the distance CQ, we shall have

$$\frac{dS}{ds} = \frac{R^2}{r^2}.$$

Also the rays sent to the lens by any point p of dS and by any point q of ds, form small right cones about the normals at p and q whose semi-vertical angles are as r to R. Hence if for simplicity we suppose dS and ds to be coated with lamp-black (so that there is no reflection), we see that the rate at which radiation is sent by dS to ds through the lens is equal to the rate at which it is sent by ds to dS.

59. Advantage has been taken, by Sir J. Herschell and Pouillet, of the peculiar property possessed by lamp-black of absorbing practically all the radiation which falls upon it, to estimate the amount of energy radiated by the sun to the earth. Of course, it is only possible to measure the quantity that actually reaches the surface of the earth; but the amount prevented from reaching the surface by the atmosphere can be roughly allowed for by taking observations when the sun has different altitudes, and

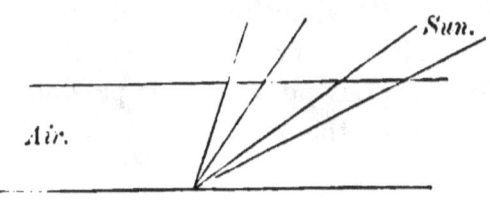

when, in consequence, different thicknesses of air have to be traversed, as shown in the figure. It thus appears probable that only one half of the radiant energy sent by the sun to the earth succeeds in reaching the surface. Taking this into account, it is calculated that the quantity of radiation which falls on the earth in a year would be sufficient to melt a layer of ice 30 metres (100 feet) thick, covering the entire surface.

Since a gramme of ice at 0° C. occupies 1·087 cubic centimetres, and since the latent heat of ice is 79·25 calories, the total radiation which falls on the earth from the sun in a year is equivalent to about 3000×73 calories, or 9×10^{12} ergs, per square centimetre. This is about the same as 25 calories, or 10^9 ergs, per square centimetre per hour; or again, 3×10^5 ergs ($\frac{1}{327}$ kilogrammetre) per square centimetre per second. In English measure, the average quantity of radiation per square foot which actually reaches the surface of the earth in an hour, is sufficient to raise the temperature of $25\frac{1}{2}$ lbs. of water from 0° C. to 1° C., which is equivalent to 35,500 foot-pounds.

Now if a particle receives a small displacement of ds centimetres whilst it is acted on by a force whose resolved part in the direction of the displacement is p dynes, the work done by the force on the particle is pds ergs. Consequently, if the time of describing the elementary path ds be dt seconds, the rate at which work is done by the force on the particle is $p\dfrac{ds}{dt}$, or pv, ergs per second. If therefore we divide the rate at which work is being done by the velocity of the particle, we shall get the resolved part in the direction of motion of the force which acts upon it. Suppose then, merely for the purpose of the following illustrative numerical calculation, that each particle of a body possesses the same non-mechanical velocity and let the mechanical motions be so small that they need not be taken into account. Then if we divide the rate at which radiation is being gained by the common non-mechanical velocity of each particle, we shall obtain the sum of the resolved parts in the direction of motion of the radiation forces which act on the several particles, or, as we may call it, the total radiation force acting on the body. Thus if we take the non-mechanical velocity of each particle of the earth's crust to be 50,000 centimetres per second, and suppose the total gain of radiation by the crust to be x times the radiation which actually reaches it, the total radiation force per square centimetre will be $3x$ dynes; which is about the same as $30\cdot 6x$ grammes per square metre. In English measure, the total radiation force per square foot would be about $\dfrac{x}{160}$ lb.

Of course, the non-mechanical motions really take place in such different directions that, in the case of any finite area, the resultant of all the radiation forces, found by the ordinary rules of statics, will generally be quite insignificant. When a material system has attained its final state, as explained in Art. 56, the resultant will be strictly zero; but when the temperatures of different parts of a body are unequal and undergoing rapid variations, this is not necessarily the case. Thus, for example, the orbit of a comet is generally in the form of a very elongated ellipse with the sun in one focus, and when the comet is describing that part of the orbit nearest the sun, its velocity becomes very great. Hence, since the intensity of solar radiation evidently varies inversely as the square of the distance from the sun's centre, as the comet rushes close past the sun, the intensity of the radiation which falls upon it will increase with enormous rapidity to a maximum and then decrease as fast. Under these conditions, it would scarcely be possible for the radiation forces which act on the comet to be so delicately balanced as to produce no visible mechanical effect. Perhaps the tails of comets may be due to this cause; for they appear to be produced by a repulsive force residing in the sun, and we know that the force required is generally small, the mass of a comet's tail being sometimes under 100 grammes (3·5 ounces).

If a be the radius of the earth and d its distance from the sun, the total energy radiated by the sun is $\frac{4\pi d^2}{\pi a^2}$ times as much as falls on the earth in the same time. Since $\frac{d}{a} = 23000$, we find $\frac{4\pi d^2}{\pi a^2} = 2100$ million. Hence if b be the radius of the sun, the radiation emitted by the sun per hour is sufficient to melt a layer of ice covering its entire surface $21 \times 10^8 \times \frac{4\pi a^2}{4\pi b^2} \times \frac{30}{365 \times 24}$ metres thick.

Substituting 110 for $\dfrac{b}{a}$, we find this thickness to be about 600 metres (660 yards). Again, the radiation from a given area of the sun is $21 \times 10^8 \times \dfrac{4\pi a^2}{4\pi b^2} \equiv 175000$, times as much as falls on an equal area of the earth in the same time. It is therefore equal to 4,375,000 calories per square centimetre per hour; or to 1215 calories, or 525×10^8 ergs, per square centimetre per second. In English measure, the radiation per hour for each square foot is sufficient to raise the temperature of 90,000 lbs. of water from $0°$ C. to $100°$ C.

If distances be expressed in centimetres, we have $a = 64 \times 10^7$, and therefore the total radiation emitted by the sun in a year is

$$(21 \times 10^8) \times (4\pi a^2 \times 9 \times 10^{12}) = 10^{41} \text{ ergs.}$$

60. It may be taken that the furnace of a locomotive consumes 1 lb. of coal per minute for each square foot of grate, and that each pound of coal burnt gives out sufficient heat to raise the temperature of 75 lbs. of water from $0°$ C. to $100°$ C. The radiation from a given area of the sun is therefore 20 times the heat generated in the same time in a locomotive furnace with the same area of grate. In making the comparison, it must, however, be borne in mind that while the furnace of a locomotive is of very small depth, the radiation emitted by the sun is not a mere surface phenomenon, but probably comes to a large extent from considerable distances below. In fact, while the absolute temperature of the fire of a locomotive may be as high as $1800°$, the temperature of the sun's surface is only estimated at from $3000°$ to

10000°. The central parts of the sun will, of course, be at a higher temperature.

61. Proceeding now to a very important extension of Carnot's axiom, let us take any material system which may possess any mechanical motions and may be electrified in any way we please, but is not subjected to external electric influences; and suppose it made to undergo a cyclical process during which it is at liberty to lose or gain heat in any possible way. Whilst the temperature θ at any point P of the system varies continuously by an infinitesimal amount $d\theta$ in the time dt, let the small quantity of heat *absorbed* near the point P, either by conduction or radiation, be denoted by $dt\,dh$. Whether the point P be on the surface or far in the interior, the heat $dt\,dh$ may be supposed, by a stretch of the imagination, to be supplied by contact with a vessel filled with a liquid and its vapour. If the liquid be sufficiently volatile, the temperature will be uniform throughout the vessel; for a difference of temperature in any part would cause an explosive formation or condensation of vapour until the temperature became uniform. Also if the sides of the vessel be supposed to be very good conductors and to have an exceedingly small capacity for heat, the temperature of the liquid, when the vessel is in contact with any part of the system, will be the same as the temperature of that part of the system with which the vessel is in contact. We are therefore at liberty to suppose that all the heat $dt\,dh$ is imparted by a 'Carnot's perfectly reversible engine' which continually brings it from a source whose temperature is uniform and constantly equal to θ_0. The heat dq_0 taken from the source to supply a quantity of heat $dt\,dh$ at the temperature θ being given by the relation

$$\frac{dq_0}{\theta_0} = \frac{dt\,dh}{\theta},$$

we have
$$dq_0 = \theta_0 dt\,\frac{dh}{\theta}.$$

The total amount of heat taken from the source in the time dt will therefore be

$$\theta_0\,dt \int \frac{dh}{\theta},$$

where the integral extends throughout the system at the time t. In a complete cycle, the total quantity of heat taken from the source will be

$$\theta_0 \int dt \int \frac{dh}{\theta}.$$

Hence, since in any complete cycle heat cannot be taken from a body whose temperature is uniform and constant unless some other body of different temperature be also present, and since θ_0 is always positive, we infer that, for a non-frictional cycle, the integral $\int dt \int \frac{dh}{\theta}$ is zero, and that for all other cycles it is negative.

If the temperature of the system be uniform both at the beginning and at the end of the cycle, these results may be more simply expressed by saying that the integral $\int \frac{dQ}{\theta}$ is zero for non-frictional cycles and negative in all other cases, dQ being the heat absorbed by the whole system whilst its temperature varies continuously from θ to $\theta + d\theta$, which may take place at different times in different parts of the system.

62. The results of the preceding article lead to some important conclusions relating to those states of a system through which it can be made to pass by non-frictional methods; for if P, Q be any two such states, it may be assumed, as a result of observation and experience, that the system can be made to pass from P to Q by a non-frictional process. Thus if a system pass from an initial state O to a second state A by any non-frictional path OPA, and then return from A to O by a non-frictional path AXO, and the corresponding integrals be distinguished by suffixes, as below, we have

$$\int dt \int \frac{dh}{\theta}_{OPA} + \int dt \int \frac{dh}{\theta}_{AXO} = 0.$$

Since the system in the states O and A is either of uniform temperature throughout or consists of a number of different bodies of different uniform temperatures, the last equation may also be written

$$\Sigma \int \frac{dQ}{\theta}_{OPA} + \Sigma \int \frac{dQ}{\theta}_{AXO} = 0,$$

where θ now denotes the uniform temperature of any one of the different bodies at any instant during the process.

If the system pass from O to A by another non-frictional path OQA, and then return by the same path AXO, as before, we have

$$\int dt \int \frac{dh}{\theta}_{OQA} + \int dt \int \frac{dh}{\theta}_{AXO} = 0,$$

or
$$\Sigma \int \frac{dQ}{\theta}_{OQA} + \Sigma \int \frac{dQ}{\theta}_{AXO} = 0.$$

Hence

$$\int dt \int \frac{dh}{\theta}_{OPA} = \int dt \int \frac{dh}{\theta}_{OQA} \quad \ldots\ldots\ldots\ldots(33),$$

CARNOT'S PRINCIPLE.

which is the same as

$$\Sigma \int \frac{dQ}{\theta}_{OPA} = \Sigma \int \frac{dQ}{\theta}_{OQA} \quad \ldots \ldots \ldots \ldots (33)'.$$

These important equations were given by Sir W. Thomson and Clausius about the same time. In the hands of Clausius they have led to some further results which may be classed as the most remarkable in the whole range of thermo-dynamics. For since the integrals just obtained depend only on the two states O, A, if the state O be chosen as a standard fixed state, each integral will depend only on the state A, and may be written ϕ_A. The function ϕ was called by Clausius the Entropy of the system; and since its value would only be altered by a constant if a different state were taken as the standard state, it is clear that the entropy is a single-valued function of the independent variables which define the state of the system together with an arbitrary additive constant depending only on the choice of the standard state.

If the system pass in a non-frictional manner from O through A to another state B, we obtain

$$\int dt \int \frac{dh}{\theta}_{OAB} = \phi_B,$$

or
$$\int dt \int \frac{dh}{\theta}_{OA} + \int dt \int \frac{dh}{\theta}_{AB} = \phi_B,$$

or
$$\int dt \int \frac{dh}{\theta}_{AB} = \phi_B - \phi_A \quad \ldots \ldots \ldots \ldots (34),$$

which may also be written

$$\Sigma \int \frac{dQ}{\theta}_{AB} = \phi_B - \phi_A \quad \ldots \ldots \ldots \ldots (34)'.$$

Hence if the system be prevented from receiving or losing heat, all the different states into which it can be brought by non-frictional processes have the same entropy. Such non-frictional processes are called Adiabatic or Isentropic processes.

Again, if the temperature of every part of the system keep the same constant value θ, $(34)'$ takes the simpler form

$$\Sigma \int dQ_{AB} = \theta (\phi_B - \phi_A) \ldots\ldots\ldots\ldots(34)''.$$

When the path AB is frictional, let the system return from B to A by a non-frictional path, so as to complete the cycle. Then by the last article and by equation (34), we have

$$\int dt \int \frac{dh}{\theta}_{AB} + \phi_A - \phi_B < 0,$$

that is,
$$\int dt \int \frac{dh}{\theta}_{AB} < \phi_B - \phi_A \ldots\ldots\ldots\ldots(35).$$

Since the system in the states A, B consists of a number of bodies of uniform temperatures, we obtain

$$\Sigma \int \frac{dQ}{\theta}_{AB} < \phi_B - \phi_A \ldots\ldots\ldots\ldots(35)',$$

where dQ is the heat absorbed by any part of the system while its temperature varies continuously from θ to $\theta + d\theta$.

As no process in nature can exactly be a non-frictional process, it follows from equation (35) or $(35)'$, that when a system is prevented from losing or gaining heat, its entropy must be continually increasing.

If the definition of entropy apply to two states in-

CARNOT'S PRINCIPLE. 143

definitely near together, then when the system passes from one to the other, we have

$$dt \int \frac{dh}{\theta} \lesseqgtr d\phi \quad \text{...................(36)},$$

or

$$\Sigma \frac{dQ}{\theta} \lesseqgtr d\phi. \quad \text{...................(36)}'.$$

63. The simple relation $\Sigma \dfrac{dQ}{\theta} = d\phi$, which holds for any small non-frictional change of state, becomes $dQ = \theta d\phi$ when we consider only a single body of uniform temperature. Hence if we take two rectangular axes and denote the temperature and entropy of the body at any instant by

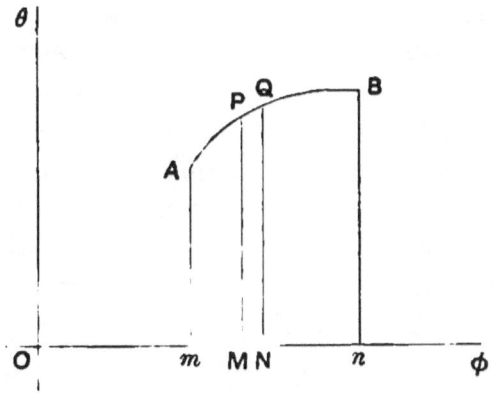

the ordinate PM and abscissa OM of a point P, the heat absorbed by the body when a small change of state PQ is produced in a non-frictional manner will be represented by the small area $PQNM$. Consequently in a finite non-frictional operation AB, the heat absorbed, $\int_A^B \theta d\phi$, is represented by the area of the figure $ABnm$, and therefore generally depends on the form of the curve AB as well as on the initial and final states A, B: in other words, dQ is not generally a complete differential of a function of the independent variables which define the state of the system. The heat absorbed during the process AB will be the same for all non-frictional paths leading from A to B, or dQ a complete differential at every point of the path, under any one of the three following conditions:—

(1) When the temperature is constant throughout every path; or

(2) When the entropy is constant; or

(3) When the temperature is a function of the entropy.

64. The conception of entropy enables us to remove an objection which is sometimes urged against Carnot's principle. It is commonly stated that an animal has a much greater 'efficiency' than a Carnot's perfectly reversible engine working through the same range of temperature, and it is even compared to an electromagnetic engine, although it is obvious that there is no sensible absorption of electric energy. In consequence it is believed that the changes which take place in the animal world are not subject to the restrictions of Carnot's principle. The whole difficulty, however, arises from the fact that the word 'efficiency' is unconsciously used in a new sense and that it is supposed it should then possess the same properties as when used strictly according to the definition. In Carnot's principle, the 'efficiency' is the ratio of the work done during a *complete* cycle to the mechanical value of the heat absorbed from the hotter body. In the present example, the 'efficiency' is the ratio of the work done in an *incomplete* cycle to the mechanical value of the consequent change of energy. If, for example, a system undergo a reversible operation, at the constant temperature θ, during which the energy and entropy change from U_1 and ϕ_1 to U_2 and ϕ_2, respectively, the heat given out by the system will, by equation (34)″, be equal to $\theta(\phi_1 - \phi_2)$, and therefore the work done by the system will be $U_1 - U_2 - \theta(\phi_1 - \phi_2)$. Thus the ratio of the work done to the mechanical value of the loss of energy is

$$\frac{U_1 - U_2 - \theta(\phi_1 - \phi_2)}{U_1 - U_2}, \text{ that is, } 1 - \theta\frac{\phi_1 - \phi_2}{U_1 - U_2},$$

which, when no sensible amount of heat is lost or gained, will be practically equal to unity.

The cycle of the animal body is completed in the vegetable world, and though very little is really known at present of the way in which the growth of grass and trees takes place, we have little doubt that when the deficiency is supplied, every part of the cycle will be found to be in complete accordance with Carnot's principle.

65. We have already seen (Art. 25) that when a body is in a state to which the definition of entropy applies, the velocity of its centre of mass may be varied at will by an isentropic process without altering the temperature or any other property of the body. Hence when two different states of the same body or system of bodies differ only in the motion of the centre of mass, the conception of entropy will be applicable to both states or to neither; and in the former case, the entropy will have the same value in both states, so that it is independent of the motion of the centre of mass.

If the speed of rotation of a body be altered by an isentropic process, the temperature and other properties of the body will generally vary in consequence; but if the body be made of very rigid materials, like a fly-wheel, these variations will generally be extremely small. The entropy of a rigid body, like a fly-wheel, is therefore very nearly independent of the speed at which it is rotating.

66. If we take the solar system for our material system, the conception of entropy will not be applicable, on account of the irreversible frictional actions which are continually taking place; but it will be possible, at any instant, by suitable means, without doing work on the system, or allowing it to receive or lose heat, to bring it into a slightly different state to which the definition of entropy applies. To do this, it will be practically sufficient to stop the mechanical and thermal irreversible frictional actions by holding the tides rigidly in their places, and preventing the conduction and radiation of heat between parts of the sun's mass which differ much in temperature by enclosing them in good non-conducting materials able to withstand the fiercest temperatures to which they may be exposed. Suppose, then, that P is the actual state of the solar system at any instant and Q the actual state at any subsequent instant, and that P', Q' are two other states to which the conception of entropy applies and which differ very little from P, Q, respectively. Then since the radiation of energy into infinite space is the only external influence to which the solar system is subject, it is obvious that the energy is less in the state Q than in the state P; and that we may bring the system from the state P' to the state Q' by a reversible process in which a positive quantity of heat is lost, so that the entropy in the state Q' is less than in the state P'. We may therefore say that the energy and entropy of the solar system are both continually decreasing.

The variation of the energy and entropy of the solar system is not considered in the ordinary books; but we frequently meet with the discussion of a similar problem relating to the whole universe which seems to require notice. In the first place, the discussion in question ignores the all-important distinction between bound etherial energy, that is, the potential energy of matter, and free etherial, or radiant energy, by classing the whole of the ether with material bodies. It is consequently concluded that the energy of the universe is constant. In the second place, it is supposed that in the equation

$$\int dt \int \frac{dh}{\theta} < \phi_B - \phi_A$$

we may take $dt\,dh$ to be an element of the radiant heat which enters at the external boundary of the ether, and not an element of the heat absorbed by some material part of the system, as in the proof of the formula given in Art. 62. With this assumption, every element $dt\,dh$ is zero and therefore the whole integral $\int dt \int \frac{dh}{\theta}$ is also zero, so that it would follow that the entropy of the universe continually increases.

In the present state of our knowledge, it is perhaps unsafe to make any positive assertion as to the energy or entropy of the whole material universe.

67. We will now consider the important thermodynamical problem of the capacity of a material system

148   ELEMENTARY THERMODYNAMICS.

for doing mechanical work when it is subject to certain simple restrictions as to the way in which it receives or loses heat. The only cases that will be here investigated are the following:—

(1) The system is supposed to be entirely prevented from receiving or losing heat, either by conduction or radiation; and the maximum amount of work that can be obtained from the system under these conditions is defined to be the Adiabatic Available Energy.

(2) The system is supposed to be prevented from radiating energy into infinite space and from exchanging heat with external bodies *except* with bodies A, B, ..., which are kept at any the same constant uniform temperature θ. For this purpose it will generally be necessary to employ an intermediate body A', just as when we explained Carnot's axiom in Art. 48. Under these conditions, the maximum amount of work that can be obtained from the system is defined to be the Available Energy at the constant temperature θ.

In both cases, the algebraic excess of the available energy over the mechanical kinetic energy is defined to be the Mechanical Potential Energy. Hence since there are two kinds of mechanical potential energy, it follows that both kinds cannot be identical with true potential energy. This result will perhaps be made clearer if we repeat some of our former conclusions relating to energy.

According to modern ideas, all actions at a distance between particles of matter are due to continuous contact forces in the ether. Consequently, all energy is really kinetic energy, and consists partly of the kinetic energy of matter and partly of the kinetic energy of the ether. The kinetic energy of matter may be distinguished as mechanical or non-mechanical, kinetic energy. The kinetic energy of the ether may also be divided into two classes:—

(*a*) Etherial kinetic energy free from all material restraints, as in waves of radiation.

(*b*) Etherial kinetic energy bound to material systems and forming the true potential energy of matter.

There is no potential energy in the ether, because that would require actions at a distance between the particles of ether, which it would need a second kind of ether to explain[1].

Now according to the kinetic theory of gases, which is not explained in this book, the pressure of a gas is due to the fact that its 'molecules' are moving about freely among one another in all directions with great velocities. The mechanical potential energy of a compressed gas therefore consists, to a large extent, of the kinetic energy of matter. In the case of a bent spring it is not so obvious at first whether the mechanical potential energy consists of kinetic

---

[1] It may be here pointed out that quite a different meaning is assigned to the term potential energy in ordinary dynamics from that which is given to it in thermodynamics. In ordinary dynamics, when the expression for the *external* work in a small change of state is a complete differential, the potential energy of the system is defined to be the work which the *external forces* can do on the system as it returns from its actual state to the standard state. Thus when a stone is lifted from the ground, it is usually said to acquire potential energy. In thermodynamics, if the stone be considered as a complete system in itself, it is said to acquire no potential energy; but if we suppose it to form part of one system with the earth, then the system acquires additional mutual potential energy. But we cannot assert that this mutual potential energy belongs necessarily to the stone; for if the stone fall to the earth, the earth rises to meet the stone, so that the kinetic energy, which takes the place of the mutual potential energy, is divided between them.

or potential, energy; that is, whether the mechanical potential energy may not be due to actions at a distance. But even in this case, there is little doubt it will be found to consist, to a large extent, of the kinetic energy of matter. In the vibrations of a spring we therefore see the transformation of non-mechanical into mechanical, kinetic energy, and conversely.

68. When a material system has been deprived of its available energy, it is evident that its mechanical motions, both of translation and rotation, will be zero, and that the temperatures of all its parts will be equal. The ultimate electric and magnetic condition of the system will not be discussed here, but may be understood from works on electricity. As to the relative positions of its parts, it is obvious that when any portion of the system is gaseous, it may be necessary to impose some minor restrictions in addition to those which specify that the system is not to receive or lose heat, or only to exchange heat with bodies A, B, ..., which are kept at the same constant uniform temperature. For instance, if the gas be contained in a closed vessel, we may impose the condition that the vessel is to be kept gas-tight; if the vessel be in the form of a cylinder fitted with a gas-tight piston, we may suppose that the expansion is not to proceed beyond a certain point, or that the final pressure is to have a given value. Even when the system is entirely solid or liquid, we may have minor restrictions; for it may be required that all parts of the system preserve their individual identity, or we may be at liberty to weld them all into one homogeneous mass. The minor restrictions, it may be remarked, are generally understood without being expressly stated, and are not to prevent the ultimate exchange of heat between different parts of the system or between the system and the bodies A, B, ...

69. In order to apply the principle of reversibility to the investigation of available energy, it is necessary to restrict ourselves to those states of the system to which the conception of entropy applies.

In the first place, let the system be prevented from receiving or losing heat, and suppose that it is brought by a reversible path from an initial state P, to which the conception of entropy is applicable, to a final state Z, in which the available energy is zero. Then the entropy is the same in the state Z as in the state P; and it may be assumed, as a result of experience, that if the system be subject to the same minor restrictions as before, and be brought by a different reversible path from the initial state P to a final state Y, the states Y and Z will be identical. Let these paths be represented by (1) and (2), and denote the corresponding amounts of work obtained from the system by W_1 and W_2. Then if the system be made to pass from P to Z by (1), and then to return from Z to P by (2), it will have undergone a complete reversible cycle in which there is neither loss nor gain of heat. The work $W_1 - W_2$ obtained from the system must therefore be zero, or $W_1 = W_2$.

If the system be brought from P by a frictional path to a final state Z', in which the available energy is zero, the state Z' cannot be the same as Z; because if ϕ be the entropy in the state P, the entropy in the state Z will be equal to ϕ, and in the state Z', greater than ϕ. The system, however, will evidently admit, in general, of being brought from Z to Z' by a frictional path in which a positive quantity of work is expended (in friction). Suppose, then, that W' is the work obtained from the frictional path PZ', and W the work obtained from any reversible path PZ, and let the system be brought from P to Z' in the two following ways:—

(1) Let the system travel by the frictional path PZ' yielding a quantity of work W'.

(2) Let the system first travel from P to Z by a reversible path, giving out a quantity of work W; and then from Z to Z', giving a negative quantity of work, $-x$ say, where x is positive. Then since there is neither loss nor gain of heat in the paths leading from P to Z', the principle of energy gives
$$W' = W - x,$$
or
$$W > W'.$$

Hence if U and U_0 be the values of the energy in the states P and Z, the adiabatic available energy in the state P is $U - U_0$, with the condition $\phi = \phi_0$.

Secondly, let us suppose that the system exchanges heat with bodies which are kept at the same constant uniform temperature θ. Then if the system, subject to given minor restrictions, be brought by any operation from the state P to a final state Z, in which the available energy is zero, it is obvious that the temperature of the system will then be uniform and equal to θ, and that the state Z will be the same for all operations. From this it may be shown that the work W obtained from the system as it travels from P to Z is the same for all non-frictional paths, and is then greater than for any frictional path. Since, in a non-frictional path, no part of the system is allowed to receive or give out heat except when its temperature is θ, if (U, ϕ), (U_z, ϕ_z) be the energy and entropy of the system in the states P and Z, respectively, the heat absorbed in a reversible path leading from P to Z is $\theta(\phi_z - \phi)$, or the heat given out is $\theta(\phi - \phi_z)$, and therefore the work obtained from the path is $U - U_z - \theta(\phi - \phi_z)$. Writing $\mathcal{F}$ for $U - \theta\phi$, this becomes $\mathcal{F} - \mathcal{F}_z$, where, it must be remembered, ϕ is not equal to ϕ_z.

70. If, while a system is protected from all external influences, including as such the radiation of energy into external space, an action takes place within the system by which the state is changed from P to P', the energy will remain the same but the entropy will be increased. Thus if (U, ϕ), (U', ϕ') be the energy and entropy in the states P, P', respectively, we shall have

$$\left.\begin{array}{l} U' = U \\ \phi' > \phi \end{array}\right\}.$$

Now let W, W', be the adiabatic available energies in the states P and P', Z and Z' the corresponding final states; and let U_0, U_0', be the values of U in the final states Z, Z'. Then we have

$$\left.\begin{array}{l} W = U - U_0 \\ W' = U' - U_0' \end{array}\right\},$$

so that $\qquad W - W' = U_0' - U_0.$

But since the entropy in the final state Z' is equal to ϕ', and in the final state Z equal to ϕ, which is less than ϕ', and since there are no mechanical motions in either state, it may be inferred that the system can be brought from Z to Z' by a frictional path in which a positive quantity of work is expended (in friction) and no heat either gained or lost. We therefore have $U_0' > U_0$, so that $W > W'$.

Again, if V, V', be the available energies at the con-

stant temperature θ in the two states P, P', and if Z be the common final state,
$$\left. \begin{array}{l} V = \mathcal{F} - \mathcal{F}_z \\ V' = \mathcal{F}' - \mathcal{F}_z \end{array} \right\},$$
so that
$$\begin{aligned} V - V' &= \mathcal{F} - \mathcal{F}' \\ &= (U - \theta\phi) - (U' - \theta\phi') \\ &= \theta(\phi' - \phi), \end{aligned}$$
and therefore $V > V'$.

Since, therefore, when a system is left to itself, the entropy has a constant tendency to increase, it follows that there is a constant tendency to a loss of available energy; in other words, to a transformation of available into unavailable, energy. This result is the great principle of the Degradation of energy, first enuntiated by Sir W. Thomson. It was formerly known as the principle of the Dissipation of energy; but the use of the word dissipation is now generally abandoned, since, as we have seen, there is no actual loss of energy when the system passes from P to P'.

Again, since it is obvious that when a finite system is left to itself, the entropy cannot increase indefinitely, we conclude that the system will ultimately settle down into a permanent state or condition, as stated in Art. 26. The principal agency by which this condition is generally brought about will be explained a little later on; in the meantime it may be stated that the ultimate condition is not necessarily such as to make the available energy zero.

71. When we wish to obtain the maximum amount of work from a system which is prevented from radiating energy into infinite space, and also from exchanging heat with any other system, or which is only allowed to ex-

change heat with bodies which are kept at the same constant uniform temperature, we are at liberty to bring the system to its final state by any reversible process we please. If, for example, the system consists of a number of different bodies or parts, we may begin by depriving each part of its available energy. If, during the rest of the process, the state of each part can be kept invariable, we shall have Prof. Tait's proposition that the available energy of a system is equal to the sum of the available energies of each of its parts together with their mutual available energy after their individual available energies have been exhausted.

We will now take a few particular examples of adiabatic available energy for systems in which there are no manifestations of electricity or magnetism.

(1) Let the system consist of a number of bodies A together with a vast body B whose temperature is uniform and equal to θ_0 and whose volume is constant. Then since the maximum amount of work may be obtained by any reversible path we please, we will take the simplest, as follows. Let the bodies A undergo any set of isentropic operations by which the temperature of every part is reduced to θ_0, and then let them be put in thermal communication with B and bring the system to its final state by any set of isothermal operations. If (U, ϕ) be the original values and (U_0, ϕ_0) the final values, of the energy and entropy of the bodies A, the heat given out by them to B will, by equation (34)″, be equal to $\theta_0(\phi - \phi_0)$, and therefore the maximum amount of work which can be obtained from the system, or its adiabatic available energy, will be

$$U - U_0 - \theta_0(\phi - \phi_0) \quad \ldots\ldots\ldots\ldots (37).$$

Putting $\mathcal{F}$ for $U - \theta_0 \phi$, this becomes $\mathcal{F} - \mathcal{F}_0$.

In this example we have evidently found the available energy at the constant temperature θ_0, of the bodies A alone.

(2) Let the system consist of masses $m_1, m_2, m_3, \ldots m_n$ of the same kind of perfect gas contained in cylinders fitted with air-tight pistons whose volumes may be varied in any way subject to the condition that the total volume remains constant. Also let the temperatures of the different masses be originally $\theta_1, \theta_2, \theta_3, \ldots \theta_n$, and their volumes per unit mass $v_1, v_2, v_3, \ldots v_n$, respectively. Then since for a unit mass of perfect gas we have very approximately $dU = C_v d\theta$, and since the work done on the gas in a small reversible operation is $-pdv$, the equation $dU = dQ + dW$ gives for a small reversible process

$$dQ = C_v d\theta + pdv,$$

or since the gas satisfies very approximately the relation $pv = R\theta$,

$$dQ = C_v d\theta + R\theta \frac{dv}{v}.$$

Hence since
$$\frac{dQ}{\theta} = d\phi,$$

we have
$$d\phi = C_v \frac{d\theta}{\theta} + R \frac{dv}{v}.$$

Integrating and remembering that C_v is practically constant, we obtain

$$\phi - \phi' = C_v \log \frac{\theta}{\theta'} + R \log \frac{v}{v'}.$$

Thus when the given masses of gas are brought in any reversible way to a final state in which each quantity has the same pressure p_0 and the same temperature θ_0, the

CARNOT'S PRINCIPLE. 157

conditions that the total entropy and the total volume are the same as at first, are

$$C_v \left\{ m_1 \log \frac{\theta_1}{\theta_0} + m_2 \log \frac{\theta_2}{\theta_0} + \ldots \right\}$$
$$+ R \left\{ m_1 \log \frac{v_1}{v_0} + m_2 \log \frac{v_2}{v_0} + \ldots \right\} = 0,$$

or $\left(\dfrac{\theta_1{}^{m_1} \theta_2{}^{m_2} \theta_3{}^{m_3} \ldots \theta_n{}^{m_n}}{\theta_0{}^{\Sigma m}} \right)^{C_v} \times \left(\dfrac{v_1{}^{m_1} v_2{}^{m_2} v_3{}^{m_3} \ldots v_n{}^{m_n}}{v_0{}^{\Sigma m}} \right)^{R} = 1,$

and $v_0 \Sigma m = V.$

Hence

$$\theta_0{}^{C_v \Sigma m} = \left(\frac{\Sigma m}{V} \right)^{R \Sigma m} \times (\theta_1{}^{m_1} \theta_2{}^{m_2} \ldots \theta_n{}^{m_n})^{C_v}$$
$$\times (v_1{}^{m_1} v_2{}^{m_2} \ldots v_n{}^{m_n}) \ldots \ldots \ldots (38),$$

which gives θ_0. To find p_0, we use the equation

$$p_0 V = R \theta_0 \Sigma m.$$

The equation $dU = C_v d\theta$, or $U - U' = C_v(\theta - \theta')$, then shows that the work obtained from the system is

$$C_v (\Sigma m \theta - \theta_0 \Sigma m) \ldots \ldots \ldots (39).$$

If the masses $m_1, m_2, m_3, \ldots m_n$, are each equal to m, equation (38) becomes

$$\theta_0{}^{\frac{C_v}{R}} = \frac{nm}{V} \times (\theta_1 \theta_2 \theta_3 \ldots \theta_n)^{\frac{C_v}{nR}} \times (v_1 v_2 v_3 \ldots v_n)^{\frac{1}{n}}.$$

The foregoing examples are taken substantially from Maxwell's Theory of Heat. The following is due to Sir W. Thomson, but has been somewhat modified by Prof. Tait.

(3) Let there be two equal homogeneous bodies of masses m, of the same kind of substance, without mechanical kinetic energy but at different uniform tem-

peratures θ_1, θ_2, and suppose that the alterations of shape and volume can be neglected. Then no work can be obtained from the system, and we shall therefore suppose that we are at liberty to equalize the temperatures of the two bodies by means of a Carnot's perfectly reversible engine which performs complete cycles. Also, for simplicity, we shall take the specific heat to be independent of the temperature; and in making the calculations, the Carnot's engine may be supposed to work between the two given bodies and any third body whose temperature is kept at a uniform constant value θ_0, provided that, on the whole, no heat is thus imparted to, or abstracted from, the third body. The temperature θ_0 will evidently be the final temperature of the two given bodies.

If, then, the temperature of one of the given masses be increased by $d\theta$, the heat imparted to it will be $mcd\theta$, and if dq_0 be the corresponding quantity of heat taken from the third body, we have, by the definition given in equation (29),

$$\frac{mcd\theta}{\theta} = \frac{dq_0}{\theta_0}.$$

Thus the condition that, on the whole, heat is neither gained nor lost by the third body, gives

$$\int_{\theta_1}^{\theta_0} \frac{d\theta}{\theta} + \int_{\theta_2}^{\theta_0} \frac{d\theta}{\theta} = 0,$$

that is,
$$\log \frac{\theta_0}{\theta_1} + \log \frac{\theta_0}{\theta_2} = 0 \quad \dots\dots\dots\dots(40),$$

or
$$\theta_0^2 = \theta_1 \theta_2,$$

whence we find the final temperature,

$$\theta_0 = \sqrt{\theta_1 \theta_2} \quad \dots\dots\dots\dots\dots(41).$$

The same result might have been obtained from the consideration that the entropy remains constant.

Again, in the cycle in which the engine takes a quantity of heat dq_0 from the third body and gives a quantity $mcd\theta$ to the other, the work done on the engine is evidently
$$mcd\theta - dq_0,$$
or
$$mcd\theta\left(1 - \frac{\theta_0}{\theta}\right).$$

The maximum amount of work, W, which can be obtained from the two bodies by means of the Carnot's engine will therefore be
$$W = mc\left\{\int_{\theta_1}^{\theta_0} d\theta\left(\frac{\theta_0}{\theta} - 1\right) + \int_{\theta_2}^{\theta_0} d\theta\left(\frac{\theta_0}{\theta} - 1\right)\right\}$$
that is, by equation (40),
$$W = -mc\left\{\int_{\theta_1}^{\theta_0} d\theta + \int_{\theta_2}^{\theta_0} d\theta\right\}$$
$$= mc(\theta_1 + \theta_2 - 2\theta_0)$$
$$= mc(\theta_1 + \theta_2 - 2\sqrt{\theta_1\theta_2})$$
$$= mc(\sqrt{\theta_1} - \sqrt{\theta_2})^2 \quad \ldots\ldots\ldots\ldots\ldots(42).$$

More generally, if there be n equal homogeneous bodies of the same kind of matter, without mechanical kinetic energy but at different uniform temperatures; then if the specific heat be independent of the temperature, we shall have
$$\int_{\theta_0}^{\theta_1} \frac{d\theta}{\theta} + \int_{\theta_0}^{\theta_2} \frac{d\theta}{\theta} + \ldots\ldots = 0,$$
that is,
$$\log\frac{\theta_1}{\theta_0} + \log\frac{\theta_2}{\theta_0} + \ldots\ldots = 0,$$
or
$$\theta_1\theta_2\theta_3\ldots\ldots\theta_n = \theta_0^n,$$
and therefore
$$\theta_0 = \sqrt[n]{\theta_1\theta_2\theta_3\ldots\ldots\theta_n} \quad \ldots\ldots\ldots\ldots\ldots(43).$$

Also $W = mc \left(\int_{\theta_0}^{\theta_1} d\theta + \int_{\theta_0}^{\theta_2} d\theta + \ldots \right)$

$= mc (\theta_1 + \theta_2 + \theta_3 + \ldots + \theta_n - n\theta_0)$

$= nmc \left(\dfrac{\theta_1 + \theta_2 + \theta_3 + \ldots + \theta_n}{n} - \sqrt[n]{\theta_1 \theta_2 \theta_3 \ldots \theta_n} \right) \ldots (44).$

The adiabatic available energy of the bodies is therefore proportional to the excess of the Arithmetic mean over the Geometric mean, of their absolute temperatures.

More generally still, if there be any number of unequal bodies, without mechanical kinetic energy, the temperature of each being uniform and the specific heats independent of the temperature, we shall have

$$m_1 c_1 \int_{\theta_0}^{\theta_1} \dfrac{d\theta}{\theta} + m_2 c_2 \int_{\theta_0}^{\theta_2} \dfrac{d\theta}{\theta} + \ldots = 0,$$

that is, $m_1 c_1 \log \dfrac{\theta_1}{\theta_0} + m_2 c_2 \log \dfrac{\theta_2}{\theta_0} + \ldots = 0,$

or $\theta_1^{m_1 c_1} \theta_2^{m_2 c_2} \theta_3^{m_3 c_3} \ldots = \theta_0^{\Sigma mc}.$

And $W = m_1 c_1 \int_{\theta_0}^{\theta_1} d\theta + m_2 c_2 \int_{\theta_0}^{\theta_2} d\theta + \ldots$

$= \Sigma mc\theta - \theta_0 \Sigma m.$

72. The results which have now been given in this and in the first chapter will be more completely explained by considering some of the chief mechanical and thermodynamical questions arising out of the problem of the solar system. This problem is based on the following empirical laws, known as the Three Laws of Kepler:—

I. The planets describe elliptic orbits about the sun, which remains fixed in one focus.

II. As any planet moves in its orbit, the straight

line which joins it to the sun describes equal areas in equal times.

III. The squares of the periodic times of any two planets are to one another as the cubes of their mean distances from the sun.

These laws, published in 1619, were deduced by Kepler by an immense amount of arithmetical labour from the observations of Tycho Brahe. They are not quite exact, as we shall see presently. Hence, as the magnitudes of the sun and the planets are small compared with their distances from one another, it will be sufficient to begin by treating them as mere particles.

73. Kepler's laws lead immediately to Newton's great principle of universal gravitation, first published in 1687. The steps of the argument may be briefly presented as follows:

(*a*) Since no planet moves with uniform velocity in a straight line, it follows that every planet is acted on by a force. Also since the orbit of every planet is plane, the acceleration at right angles to the plane of the orbit is constantly zero. The resultant force on the planet therefore acts in the plane of the orbit.

(*b*) If S be the position of the sun, P, Q the positions

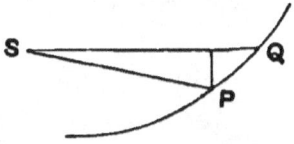

of a planet at any two consecutive instants, then if p be the perpendicular from S on the tangent at P and the elementary arc PQ be denoted by ds, the elementary area PSQ will be equal to $\tfrac{1}{2}p\,ds$.

Hence if $\dfrac{dA}{dt}$ be the rate of description of area by the line SP, we have
$$p\frac{ds}{dt} = 2\frac{dA}{dt},$$
or, if v be the velocity of the planet in its orbit,
$$pv = 2\frac{dA}{dt} \quad \ldots\ldots\ldots\ldots\ldots\ldots(45).$$

Multiplying both sides by P, the mass of the planet in grammes, we get
$$pPv = 2P\frac{dA}{dt}.$$

Now the left-hand side is the planet's moment of momentum about a straight line through S at right angles to the orbit, and by Kepler's second law, the right-hand side is constant. But we have already seen that the rate at which the angular momentum of a particle about any fixed line increases with the time is equal to the moment of the resultant force which acts upon it. The moment of the resultant force in our case is therefore zero. Consequently, the force which acts on the planet tends to or from S. Since, by Kepler's first law, the orbit is always concave to S, it is easily seen that the force always acts from P to S; in other words, that the planet is always attracted by the sun.

(c) If F be the resultant force in dynes which acts on the planet P, considered positive when it tends to the sun, and if SP, the distance of the planet from the sun, be denoted by r, the work done on the planet in a small displacement will evidently be $-F\,dr$ ergs. Hence, if

(v, r), (v_0, r_0) be the values of v and r at any two instants, the principle of work gives

$$\tfrac{1}{2} P (v^2 - v_0^2) = - \int_{r_0}^{r} F dr,$$

the integral sign merely denoting a summation, without implying that F is a function of r only.

Substituting for v from equation (45), we obtain

$$2P \left(\frac{dA}{dt}\right)^2 \left(\frac{1}{p^2} - \frac{1}{p_0^2}\right) = - \int_{r_0}^{r} F dr \quad \ldots\ldots\ldots (46).$$

Now if (p', r') denote quantities corresponding to (p, r) with respect to the other focus of the elliptic orbit, and

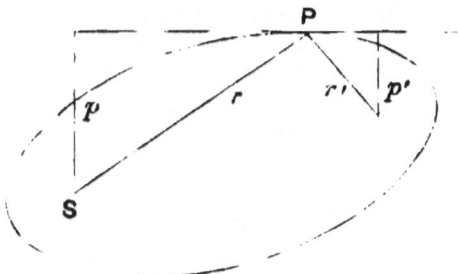

if $2a$, $2b$ be the principal axes, we have the following well-known properties:—

$$\left. \begin{array}{r} r + r' = 2a, \\ pp' = b^2, \\ \dfrac{p}{r} = \dfrac{p'}{r'} \end{array} \right\}.$$

Multiplying each side of the last of these results by p, we easily find

$$\frac{p^2}{r} = \frac{pp'}{r'} = \frac{b^2}{2a - r},$$

that is

$$\frac{1}{p^2} = \frac{1}{b^2}\left(\frac{2a}{r} - 1\right).$$

Hence by equation (46),

$$2P\left(\frac{dA}{dt}\right)^2 \frac{2a}{b^2}\left(\frac{1}{r}-\frac{1}{r_0}\right) = -\int_{r_0}^{r} F\,dr.$$

Differentiating with respect to r, and remembering that $\frac{dA}{dt}$ is constant, we find

$$F = P\left(\frac{dA}{dt}\right)^2 \frac{4a}{b^2} \cdot \frac{1}{r^2},$$

which may be more simply written

$$F = \frac{\mu P}{r^2},$$

where μ denotes the constant factor $\left(\frac{dA}{dt}\right)^2 \frac{4a}{b^2}$.

In like manner, the attractive force F', exerted by the sun on any other planet of mass P' at a distance r', may be shown to be

$$F' = \frac{\mu' P'}{r'^2},$$

where

$$\mu' = \left(\frac{dA'}{dt}\right)^2 \frac{4a'}{b'^2}.$$

(d) If T be the periodic time of the planet P, we have

$$T = \frac{\pi a b}{\frac{dA}{dt}},$$

and therefore

$$T^2 = \frac{\pi^2 a^2 b^2}{\left(\frac{dA}{dt}\right)^2} = \frac{4\pi^2 a^3}{\mu}.$$

Similarly,

$$T'^2 = \frac{4\pi^2 a'^3}{\mu'}.$$

Hence, a being the average of the greatest and least distances of the planet P from the sun, or what is usually called its mean distance, Kepler's third law gives $\mu = \mu'$. We therefore have $F = \dfrac{\mu P}{r^2}$ and $F' = \dfrac{\mu P'}{r'^2}$, so that the sum attracts the different planets with forces proportional to their masses and inversely proportional to the square of their distances.

(e) Again, if we consider only motion relative to the earth, the orbit of the moon is found to be approximately an ellipse described so that the line which joins the moon to the earth sweeps over equal areas in equal times. We therefore infer that the moon is attracted to the earth by a force $\dfrac{\nu M}{r^2}$, where M is the mass of the moon, r its distance from the earth, and ν a constant which is not to be assumed equal to μ. Now if, as a rough approximation, we were to take the moon's orbit about the earth to be circular, the force by which the moon is attracted to the earth would always be at right angles to the direction of motion, and therefore the velocity would be constant. Hence if we denote the angular velocity of the moon about the earth by ω, ω would be constant, and the resultant acceleration would tend to the centre of the orbit and be equal to $r\omega^2$. Substituting for r and ω their known values, we find (in the C. G. S. units)

$$r\omega^2 = 38 \times 10^9 \times \left(\dfrac{2\pi}{27\tfrac{1}{3} \times 24 \times 60 \times 60}\right)^2$$
$$= \cdot 27,$$

so that $\qquad \dfrac{\nu}{r^2} = \cdot 27.$

But we know that gravity, that is, the force by which the weight of bodies is caused, is proportional to the mass acted on, and at any point of the earth's surface, the force of gravity on a mass of one gramme is about 980 dynes. Hence, since the mean distance of the moon from the earth is about 60 times the earth's radius, if the force of gravity decrease inversely as the square of the distance, the acceleration towards the earth caused by gravity in any body at the distance of the moon, would be about

$$\frac{980}{60^2} = \cdot 27.$$

From this we draw the important conclusion that the force by which the moon is caused to revolve about the earth is identical with terrestrial gravity.

(f) To effect a rough comparison between μ and ν, it will be sufficient to treat the orbit of the earth about the sun as circular. Taking (r, ω) to refer to the moon's orbit about the earth, and (R, Ω) to the earth's orbit about the sun, we have

$$\left. \begin{array}{l} \dfrac{\mu}{R^2} = R\Omega^2 \\ \dfrac{\nu}{r^2} = r\omega^2 \end{array} \right\},$$

or

$$\left. \begin{array}{l} \mu = R^3 \Omega^2 \\ \nu = r^3 \omega^2 \end{array} \right\}.$$

But, roughly,

$$\left. \begin{array}{l} R = 400 r \\ \Omega = \tfrac{1}{13} \omega \end{array} \right\}:$$

hence μ is much greater than ν.

(g) From the preceding results we infer Newton's great principle of universal gravitation, which asserts that every portion of matter in the universe attracts every other portion according to the following simple law:—

CARNOT'S PRINCIPLE. 167

Let m and m' be the masses of any two portions of matter and r their distance apart, the dimensions of m and m' being small compared with r; then m and m' exert equal and opposite attractions of $\lambda \dfrac{mm'}{r^2}$ dynes on one another, where λ is a constant number which is independent of the natures of m and m', and of their chemical and physical states, and also of the presence or absence of other matter. The rigid accuracy of the law of gravitation is proved by the fact that a vast number of intricate calculations have been performed in which it has not once been found to lead us wrong. The masses m and m' may, if we please, be as small as possible, that is, in chemical language, m and m' may be the masses of atoms. We cannot, however, assert that the force of gravitation exists between the different particles of the same atom.

The universal prevalence of the attraction of gravitation has an important bearing upon thermodynamics, because it will generally produce or modify motion, and motion is generally attended by friction.

The fact that the values of μ and ν are different is easily explained by the law of gravitation. Thus if, for simplicity, we treat the sun, earth and moon, as mere particles, and if we denote their masses by S, E, M, the attractions between the sun and earth, and between the earth and moon, will be respectively equal to $\lambda \dfrac{SE}{R^2}$ and $\lambda \dfrac{EM}{r^2}$. But these forces have been denoted by $\dfrac{\mu E}{R^2}$ and $\dfrac{\nu M}{r^2}$. Hence $\mu = \lambda S$, and $\nu = \lambda E$, so that

$$\frac{\mu}{\nu} = \frac{S}{E}.$$

Since, according to Newton's law of gravitation, the planets attract the sun as much as the sun the planets, it is clear that the sun cannot remain immovable, as Kepler supposed; but owing to his great mass, his motion will be very small. Again, since the planets are not only attracted by the sun but by one another, it is evident that their orbits cannot be perfect ellipses. These questions will be found considered in detail in works on the lunar and planetary theories.

Some rough approaches to the law of gravitation had been made before the appearance of Newton's great discoveries, by considering the case of an orbit perfectly circular; but it seems that no one, with the exception, perhaps, of Horrox, had supposed that the forces by which the moon and planets are compelled to describe their orbits, are the very forces which produce terrestrial gravity. To show the connection of circular motion with Kepler's laws, let a circular orbit of radius r be described with uniform angular velocity ω by a body of mass M. Then the body will be acted on by a constant force $Mr\omega^2$ tending to the centre, and therefore if the central force be $\dfrac{\mu M}{r^2}$, we shall have

$$\mu = r^3 \omega^2.$$

Now if T be the periodic time, we have

$$T = \frac{2\pi}{\omega},$$

so that
$$T^2 = \frac{4\pi^2}{\mu} r^3.$$

Similarly, if a second body of mass M' describe another

circular orbit, of radius r', under a central force $\frac{\mu' M'}{r'^2}$, the periodic time T' will be given by

$$T'^2 = \frac{4\pi^2}{\mu} r'^3.$$

Hence $\qquad T^2 : T'^2 = r^3 : r'^3,$

which is Kepler's third law.

74. If a particle P be situated so near a finite body B that it cannot be supposed at the same distance from every part of it, it will be necessary, in order to find the resultant attraction of B on the particle P, to imagine the body B divided into a large number of parts, each of which is so small in all its dimensions that, for the purpose of finding its attraction on P, it will be sufficient to treat it as a mere particle. The simplest and most useful case to consider is that in which B is in the form of a sphere, either of uniform density throughout, or such that the strata of uniform density are shells bounded by spheres concentric with B.

Let the figure represent such a shell, of density ρ and

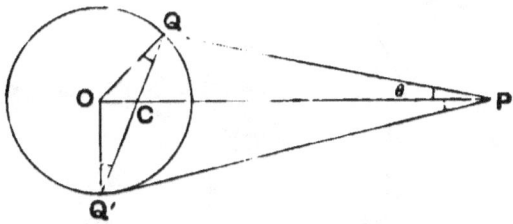

indefinitely small thickness τ. Join P to O, the centre of the shell, and take a point C in OP such that

$$OC \cdot OP = a^2,$$

where a is the radius of the shell. With C as vertex

and any small vertical solid angle $d\omega$, describe any cone cutting out a small area $d\sigma$ from the sphere at Q and another small area $d\sigma'$ at Q'. Then since OQ and OQ' make the same angle, θ say, with the line QCQ', the tangent planes to the sphere at Q and Q' will each be inclined at the angle θ to any plane normal to QCQ'. Hence
$$\left.\begin{array}{l} d\sigma \cos\theta = CQ^2 . d\omega \\ d\sigma' \cos\theta = CQ'^2 . d\omega \end{array}\right\}.$$

Again, if m be the mass of the particle P, the attractions of the elements at Q and Q' on P will be respectively equal to $\lambda \dfrac{m\rho\tau d\sigma}{QP^2}$ and $\lambda \dfrac{m\rho\tau d\sigma'}{Q'P^2}$. Substituting for $d\sigma$ and $d\sigma'$, these expressions become $\lambda \dfrac{m\rho\tau d\omega}{\cos\theta} . \dfrac{CQ^2}{QP^2}$ and

$$\lambda \dfrac{m\rho\tau d\omega}{\cos\theta} . \dfrac{CQ'^2}{Q'P^2}.$$

Now if we join the lines as in the figure, we have
$$\dfrac{OC}{OQ} = \dfrac{OQ}{OP}.$$

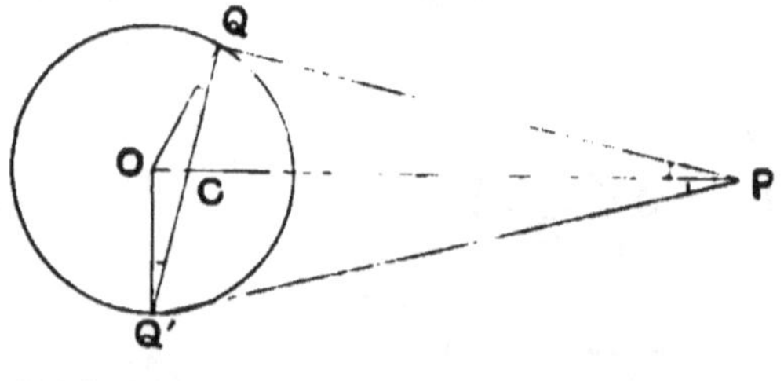

The angle OPQ is therefore equal to OQC, that is, to θ. In like manner, the angle OPQ' is equal to θ.
Also, by similar triangles, we have
$$\dfrac{OQ}{QC} = \dfrac{OP}{PQ}.$$

Therefore
$$\frac{CQ^2}{QP^2} = \frac{OQ^2}{OP^2} = \frac{a^2}{OP^2}.$$

Similarly,
$$\frac{CQ'^2}{Q'P^2} = \frac{a^2}{OP^2}.$$

Thus the attractions of $d\sigma$ and $d\sigma'$ on P are each equal to $\lambda \frac{m\rho\tau d\omega}{\cos\theta} \cdot \frac{a^2}{OP^2}$. They both lie in a plane through OP, and make the same angle θ with OP. Their resultant therefore acts along PO and is equal to

$$2\lambda \frac{m\rho\tau a^2}{OP^2} \cdot d\omega.$$

Hence also the resultant attraction of the whole shell on P acts along PO and is equal to $\lambda \frac{m \cdot 4\pi a^2 \tau \rho}{OP^2}$, that is, it is the same as if the whole shell were condensed into a particle at O. This result, being true for every such shell, is true for the whole sphere; so that if M be the mass of the sphere and r the distance OP, the attraction of the sphere on P will be

$$\lambda \frac{Mm}{r^2}.\text{[1]}$$

Now let O' be the centre of another sphere B', which, like B, is either of uniform density throughout, or such that the strata of uniform density are shells bounded by spheres which have a common centre at O'. Then in order to find the attraction of B on B', we shall first find the attraction of B on one of the thin uniform shells of which B' is composed. Let a' be the radius of such a

---

[1] The foregoing proof is from Thomson and Tait, by whom it has been modified from Newton.

172   ELEMENTARY THERMODYNAMICS.

shell, τ' its indefinitely small thickness, and ρ' its density. Join OO' and take a point C' in OO' such that

$$O'C' \cdot O'O = a'^2.$$

With C' as vertex and any small vertical solid angle $d\omega'$, describe any cone cutting out small portions from the

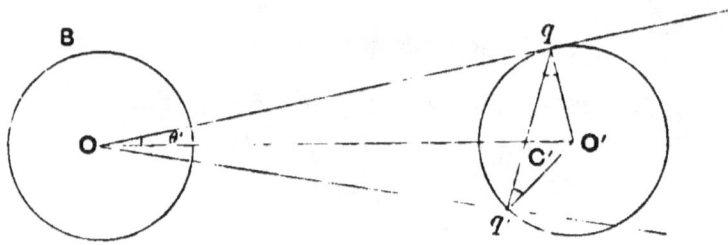

spherical shell O', at q and q'. Then it may be shown that the four angles $O'qq'$, $O'q'q$, qOO', $q'OO'$, are equal to one another, and we may therefore denote each of them by θ'. It may also be shown that $\dfrac{C'q^2}{qO^2} = \dfrac{C'q'^2}{q'O^2} = \dfrac{a'^2}{\overline{OO'}^2}.$

Again, in order to find the attraction of the sphere B on the elements at q and q', we may suppose the sphere concentrated into a point at O, and the elements at q and q' may be treated as particles. It will thence follow that, if M be the mass of the sphere B, the attractions of B on the elements at q and q' are each equal to

$$\lambda \frac{M\rho'\tau'd\omega'}{\cos\theta'} \cdot \frac{a'^2}{\overline{OO'}^2},$$

and act along qO and $q'O$, respectively.

These two forces may be compounded in the usual way. Thus, let two forces, each equal to $\lambda \dfrac{M\rho'\tau'd\omega'}{\cos\theta'} \cdot \dfrac{a'^2}{\overline{OO'}^2}$, be applied at O' in opposite directions parallel to qO. Then since the perpendicular from O' on Oq is equal to $OO' \sin\theta'$, it is easily seen that the attraction of B on the element at q is equivalent to a force $\lambda \dfrac{M\rho'\tau'd\omega'}{\cos\theta'} \cdot \dfrac{a'^2}{OO'^2}$ acting at O' parallel to qO, together with a couple $\lambda \dfrac{Ma'^2\rho'\tau'd\omega'}{OO'} \cdot \tan\theta'$ in the plane $O'qO$. In like manner, the attraction of B on the element at q' is equivalent to a force

$$\lambda \dfrac{M\rho'\tau'd\omega'}{\cos\theta'} \cdot \dfrac{a'^2}{OO'^2}$$

acting at O' parallel to $q'O$, together with a couple in the same plane as the former, and equal and opposite to it. If the different parts of the sphere B' preserve the same relative positions with respect to one another, the two opposite couples will neutralize one another (as explained in books on Statics), and need not be considered. We are therefore left with two equal forces at O' which give a resultant $2\lambda \dfrac{M\rho'a'^2}{OO'^2} d\omega'$ acting at O' along $O'O$. Integrating, we find the attraction of B on the whole shell equivalent to a single force $\lambda \dfrac{M \cdot 4\pi a'^2 \tau'\rho'}{OO'^2}$ acting at O' along $O'O$. Hence, since $4\pi a'^2\tau'\rho'$ is the mass of the shell, the result is the same as if the whole shell were condensed into a particle at O'. Treating every one of the shells of B' in the same way, we see that the attraction between the two spheres is the same as if both spheres were concentrated into particles at their centres; so that if M, M', be the masses of the spheres, and r the distance between their centres, their mutual attraction will be

$$\lambda \dfrac{MM'}{r^2}.$$

The formula which has been just obtained has two

very important applications. In the first place, it appears from the labours of astronomers, that the sun and planets are very nearly spherical bodies, each of which is composed of a number of spherical shells of uniform density concentric with one another. In treating them as if they were concentrated into points at their centres, we have therefore been making a near approximation. In the second place, an experiment known as the Cavendish experiment has been performed by Mr Baily, which virtually amounts to measuring the mutual attraction of two homogeneous spheres of known masses, from which it is calculated that the value of λ, in the C. G. S. system of units, is $\dfrac{6\cdot 48}{10^8}$. In the English system of units, in which the foot, the pound, and the poundal are units, $\lambda = \tfrac{1}{10}$. (See Routh's *Rigid Dynamics* on the Cavendish experiment; also Everett's *Units and Physical Constants*.)

It will be instructive to calculate the attraction between two equal and similar homogeneous spheres of a moderate size. Let the radius of each sphere be r centimetres, and the density ρ grammes per cubic centimetre, and let them be placed in contact. Then the mass of each sphere being $\tfrac{4}{3}\pi r^3 \rho$ grammes, and the distance between their centres $2r$ centimetres, their mutual attraction F, in dynes, will be

$$F = \lambda \frac{(\tfrac{4}{3}\pi r^3 \rho)^2}{(2r)^2}$$

$$= \lambda \tfrac{4}{9}\pi^2 r^4 \rho^2.$$

If the spheres be of lead, the value of ρ may be taken to be $11\tfrac{1}{3}$, so that the mass of each sphere $= 46\cdot 5 \times r^3$ grammes, and

$$F = \lambda \tfrac{4}{9}\pi^2 r^4 \left(\tfrac{34}{3}\right)^2$$

$$= 365 \times 10^{-7} \times r^4.$$

CARNOT'S PRINCIPLE.

By means of these formulae the following results are calculated:

Diameter of each sphere.	Mutual attraction.	Mass of each sphere.
One metre	228 dynes, or ·23 gramme, or ·0005 lb., or the 2000th of a lb.	5800 kilogrammes, or 12786 lbs., or 5·7 tons.
One centimetre	228×10^{-8} dynes, or the 200,000 millionth of a lb.	5·8 grammes, or ·013 lb.
One kilometre (1093·6 yards)	228×10^{12} dynes, or 230 million kilogrammes, or 5×10^8 lbs., or 229,000 tons.	5,800,000 million kilogrammes, or 5,700 million tons.

25·64 centimetres, or 10·1 inches	One dyne.
1·43 metres, or 56·5 inches	One gramme-weight.
2·7 metres, or 9 feet	One poundal.
6·7 metres, or 21·9 feet	One pound-weight.

75. It appears from observation that the heavenly bodies are very approximately of the forms generated by the revolution of nearly circular ellipses about their minor axes. Again, it is obvious that no body, however hard or solid, can be 'perfectly rigid,' that is, invariable as to the relative positions of its parts. For these two reasons, it is not quite accurate to treat the heavenly bodies as mere particles.

To illustrate how the want of 'perfect rigidity,' combined with the deviation from sphericity, enables gravitation to modify the celestial motions by means of friction, it will be sufficient to treat the sun and planets as if each of them consisted of two parts, (1) a rigid, homogeneous, nucleus A, in the form of the figure generated by the revolution of a nearly circular ellipse about its minor axis, and (2) an outer part, or 'tide,' B, moveable over the surface of the nucleus A. Let such a body be held so that its centre of mass is fixed at G, and so that some particular line, passing through G in a direction fixed in the body, is also fixed in space. Take the fixed direction through G as axis of z, and let a planet P, which, in its effects on the planet G, may, for the purposes of illustration, be treated as a particle, be prevented from moving except in a circular orbit about G as centre, in a plane perpendicular to Gz. Also suppose the mechanism by which these restrictions are imposed to be such that its attractive and other forces on any part of G are insignificant, except near the axis Gz, and let these forces be so symmetrical that their moment about Gz is constantly zero. Lastly, let there be no other bodies near the planet G. Then in G we practically have a body which is free to rotate about a fixed axis Gz under no exertal force but the attraction of a single particle P.

If Gz be the axis of figure of the nucleus A, it is obvious that the attraction exerted on it by P will have no moment about Gz, and that if P be fixed, the nucleus will have an infinite number of positions of equilibrium, in each of which the tide B will be symmetrical about the plane through Gz and P, being collected into heaps around the two points in which GP meets the surface of A.

If, in the first place, when P and G are in this state, we move P forward in its orbit, its attraction will carry the tide B round in the same direction; but owing to the frictional resistance which will be experienced in moving over the surface of the nucleus A, the tide will not generally continue to be symmetrical with respect to the plane zGP. On this account, the attraction of P on B will generally give a moment about Gz, and so will the frictional action exerted by B on A. The nucleus A will therefore be carried round Gz in the same direction as P; and it is clear that if the angular velocity of P in its orbit be kept constant, the nucleus A will at length acquire the same uniform angular velocity about Gz, and the tide B take up a fixed position on A symmetrical with respect to the moving plane zGP.

If, on the other hand, when G and P are in a state of equilibrium, we keep P fixed and set the nucleus A in motion about Gz, the tide B will be carried partially round with it against the attraction of P. Hence when we abandon G to the influence of P, there will be a couple about Gz by which the angular momentum of the whole body G about Gz will be ultimately reduced to zero, and the tide B brought into the same relative position with respect to zGP as before.

Again, if Gz be at right angles to the axis of figure, it may be shown that, except when the axis of figure passes through P, the attraction of P on the nucleus A will exert a moment about Gz tending to make the equator of A pass through P. When the axis of figure passes through P, the moment about Gz will be zero, but the slightest displacement from this position would bring the moment into existence. It is therefore evident that if P be at rest and the tide B absent, the nucleus A will either be at rest in a state of stable equilibrium with its equator

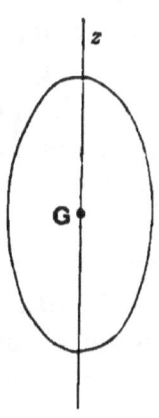

passing through P, or else will be rotating about Gz. If there be a tide B, the rotation will be checked by tidal friction, and A will finally come to rest with its equator passing through P.

In like manner, if P be rotated in its orbit with uniform angular velocity, it may be inferred that A will be carried round with it and will ultimately rotate round Gz at the same speed as P, with its equator passing through P, that is, with its axis of figure at right angles to GP.

From the preceding arguments, it is clear that if the fixed axis of rotation of the planet G be at right angles to the axis of figure, and be placed, not at right angles to the orbit of P, but in it, the planet G will ultimately place its equator in the orbit of P, or its axis of figure at right angles to that orbit, except only when P is kept at rest in either of the two points in which the axis of rotation of G produced meets its orbit. Hence also if no part of G be fixed except the centre of mass, and if P describe its circular orbit with any uniform angular velocity ω, the planet G will ultimately set its axis of figure at right angles to the orbit of P and rotate about that axis with uniform angular velocity ω.

It may therefore be inferred that, if any material system be protected from all external influences, the all-pervading force of gravitation will in time destroy all relative mechanical motions, so that the only mechanical motions of the system will then[1] consist of a constant velocity of O, the centre of mass, in a straight line, combined with a constant angular rotation of the whole system about some straight line passing through G in a fixed direction. This state, it will be easily seen, requires

[1] See Art. 24.

that the centres of mass of all the distinct bodies of which the system is composed should lie in a plane at right angles to the fixed straight line about which the whole system revolves. Furthermore, if any of these bodies be bodies of revolution, they will have their axes of figure parallel to the axis of rotation.

Now if O be the point where the axis Oz, about which the whole system rotates, meets the plane containing the centres of mass, and if G, G', be the positions of the centre

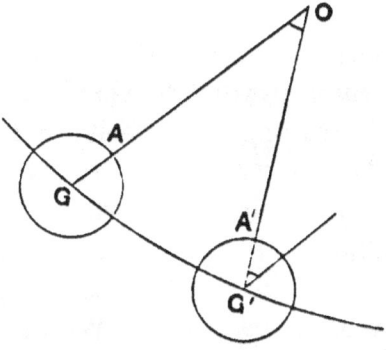

of mass of any one of the distinct bodies which compose the system, it is clear that the displacement of the body G, due to the rotation about Oz, is equivalent to a translation parallel to GG', together with a rotation of the whole body about an axis through G parallel to Oz, in the same direction as the rotation about Oz, and through an angle equal to GOG'. Hence a rotation of the body about Oz is equivalent to an equal rotation in the same direction of its centre of mass G about Oz, together with an equal rotation in the same direction of the whole body about an axis through G parallel to Oz. The final state of the system will therefore be such that:—

(1) The centres of mass of all the distinct bodies will lie in one plane, the normal to which is fixed in direction.

(2) The axes of figure of all those which are bodies of revolution will be perpendicular to this plane.

(3) In addition to a uniform velocity of all parts of the system in the same constant direction, the centres of mass will all describe circular orbits in the same direction and with a common constant[1] angular velocity, ω say, about O, the centre of mass of the whole system.

(4) Each body will also rotate about an axis through its centre of mass parallel to the fixed direction, with the same constant angular velocity ω, and in the same direction, as the centres of mass rotate about O.

The only external influence to which the solar system appears to be subject is the radiation of energy into space. This will probably continue so long as any parts are above the absolute zero of temperature. We therefore conclude that in the final state of the solar system, every part will be at the absolute zero of temperature, and consequently, that all non-mechanical motions will be absent.

The final relative positions of the bodies forming the solar system, and their final relative mechanical motions, will be of the same character as if there had been no external influences.

76. Some of the effects of friction are so important that they merit a more detailed consideration; but before we can discuss them, we need an additional property of angular momentum.

[1] See Art. 24.

Let O be any point, and suppose that OS is a straight line through O such that, at any instant t, the angular momentum of a particle P about OS is greater than about any other straight line passing through O. Draw any line OT through O making an angle θ with OS, and let it be required to compare the angular momenta of P about OS and OT at the time t.

First, let PQ, the line of motion of P at the time t be parallel to the plane SOT. Then, it is evident, PQ will lie in a plane perpendicular to OS, and will make an angle θ with a plane perpendicular to OT. Hence if p be the perpendicular distance of P from the plane SOT, m the mass of the particle P, and v its velocity, the angular momentum of P about OS will be mvp. Also if the plane through P perpendicular to OT meet OT in N, and intersect the plane through PQ parallel to SOT in

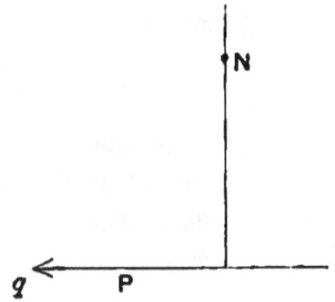

the line Pq, the perpendicular from N on Pq will be equal to p, and the resolved linear momentum of P in the plane NPq will act along Pq and be equal to $mv \cos \theta$. The angular momentum of P about OT is therefore $mv \cos \theta$.

Secondly, let the line of motion of P meet the plane SOT in R, and since the angular momentum of a particle about any line is independent of the position of the particle in its line of motion, let us suppose the particle to be at R. Draw RM and RN perpendiculars on OS and OT, and let the linear momentum of the particle at R be resolved into mv_1 at right angles to the plane SOT, and mv_2 in that plane. Then since the angular momentum

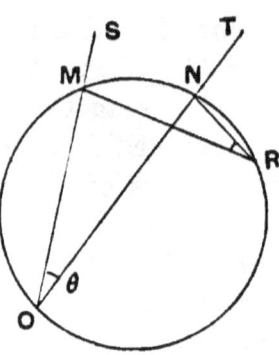

of a particle about any line is equal to the sum of the moments of its component linear momenta, the angular momentum of the particle P about OS and OT will be respectively equal to $mv_1 . RM$ and $mv_1 . RN$. But since the angles at M and N are right angles, a circle described on OR as diameter will pass through R, N, M, and O. The angle NRM is therefore equal to TOS, or θ. Also since the angular momentum about OS is greater than about any other line drawn through O, MR must be a diameter of the circle $RNMO$, or the points O and M must coincide. The angle MNR will therefore be a right angle, and $RN = RM \cos \theta$. Hence the angular momenta of the particle P about OS and OT may be written in the forms $mv_1 . RM$ and $mv_1 . RM \cos \theta$.

In both cases, if h be the augular momentum about

OS, the angular momentum about OT will be $h \cos \theta$. It thus appears that the angular momentum of P about OS is a kind of resultant, from which the angular momentum about any other line OT is to be found by resolving, exactly as if we were dealing with forces acting on a particle at O, having a resultant along OS. Consequently, if (l, m, n) be the angular momenta of any particle P about three rectangular axes Ox, Oy, Oz, and if OT be any line through O making angles (α, β, γ) with these axes, the angular momentum of the particle about OT will be

$$l \cos \alpha + m \cos \beta + n \cos \gamma.$$

Similarly, if (l', m', n') be the angular momenta of any other particle P' about the same rectangular axes, its angular momentum about OT will be

$$l' \cos \alpha + m' \cos \beta + n' \cos \gamma.$$

The angular momenta of the system of two particles about the axes and about OT are therefore respectively equal to $l + l'$, $m + m'$, $n + n'$, and

$$(l + l') \cos \alpha + (m + m') \cos \beta + (n + n') \cos \gamma;$$

whence it easily follows that if (L, M, N) be the angular momenta of a finite body about the axes, its angular momentum about OT will be

$$L \cos \alpha + M \cos \beta + N \cos \gamma.$$

Thus if we draw OA, AB, BC, parallel to the axes, and respectively proportional to (L, M, N), the angular momentum of the body about OT will be proportional to the sum of the projections of OA, AB, and BC on OT; that is, proportional to the projection of OC on OT. Consequently, if we denote $\sqrt{L^2 + M^2 + N^2}$ by H, and the angle between OC and OT by ψ, the angular momentum about

184   ELEMENTARY THERMODYNAMICS.

OT will be $H \cos \psi$. We see therefore that the angular momentum about OC is to be regarded as a resultant, and

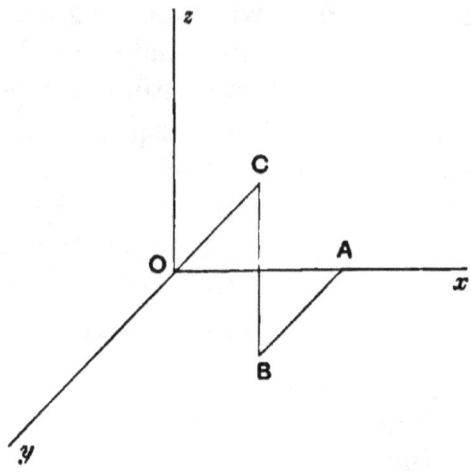

the angular momentum about any other line OT found by resolving, exactly as if we were dealing with forces acting on a particle at O and having a resultant along OC.

It should be noticed that the straight lines OS, OS', about which the angular momenta of two particles P, P', are maxima, will not generally coincide with one another, or with OC.

77. The following important propositions are collected from the first chapter, and will be required presently.

(1) The angular momentum of a body (or system of bodies), about any fixed line Oz, is equal to the angular momentum about a moveable parallel line through G, the centre of mass, together with what the angular momentum about Oz would be if the whole body (or system) were condensed into a single particle at G.

These two parts are usually referred to as the angular momenta of rotation and translation, respectively.

(2) The angular momentum of rotation depends only on the velocities relative to the centre of mass.

(3) If we assume that the internal forces of the body (or system) consist of a set of equal and opposite reactions, the rate at which the angular momentum of the body (or system) about any fixed line increases with the time, will be equal to the moment about that line of the external forces.

(4) If the internal forces consist of a set of equal and opposite reactions, the rate at which the angular momentum about a moveable line GN, through G parallel to any fixed line Oz, increases with the time, will be equal to the moment about GN of the external forces; and the rate at which the angular momentum of translation about Oz increases with the time will be equal to the moment about Oz of a system of forces applied at G equal and parallel to the external forces. Hence when there are no external forces, the angular momenta, both of translation and rotation, will remain constant with respect to any fixed line.

(5) If ω be the angular velocity of a body in which there are no mechanical vibrations, and C the moment of inertia about the axis of rotation, the angular momentum about this axis is $C\omega$.

(6) If M be the mass of the body, and V the velocity of its centre of mass, its mechanical kinetic energy will be

$$\tfrac{1}{2}MV^2 + \tfrac{1}{2}C\omega^2.$$

78. To illustrate the effects of friction and mechanical vibrations in modifying the mechanical motions of a body which is subject to no external influences, let us first take the case of an unelectrified body of the form generated by the revolution of an ellipse about its minor axis, and either homogeneous throughout, or symmetrical with respect to every plane through the axis of revolution and such that the moment of inertia about that axis is greater than about any perpendicular axis. Also let the body be extremely hard and unyielding, and the frictional effects and mechanical vibrations consequently so minute, that, for the purpose of calculating the angular momentum and mechanical kinetic energy at any instant, we may, without sensible error, suppose the relative positions of all the parts to be invariable, and represent the mechanical motions of the body by the motion of the centre of mass combined with a rotation about an axis passing through the centre of mass.

We assume, as usual, that the internal forces consist of a set of equal and opposite reactions, in consequence of which the body will possess some important properties :—

(1) The velocity V, of the centre of mass, will remain constant.

(2) The angular momentum of rotation with respect to any fixed straight line will be constant; in other words, as the body moves about, the angular momentum about a straight line drawn through the centre of mass in a fixed direction, will be invariable.

(3) Hence, as the body moves about, the resultant angular momentum at G, the centre of mass, will be constant, both in magnitude and direction.

Again, if in any state P, when the body is subject to no external influences, we apply external forces so as to bring it to a state of mechanical rest, it will be seen, on referring to Art. 25, that, on account of the hardness of the body, the mechanical work thus obtained from it may be practically equal to the mechanical kinetic energy in the state P. Suppose then that we make the body undergo the following complete cycle of operations:—

(1) Let it pass, under the action of no external influences, from any state P, in which the temperature is uniform, to any other state Q. In this operation, the entropy will increase.

(2) Let the body be then reduced to a state of mechanical rest in such a way that the mechanical work obtained is practically equal to T_Q, the mechanical kinetic energy in the state Q. Also since the temperature in the state Q will not be uniform, let sufficient time be allowed for it to become so before proceeding to the next operation.

On account of the irreversible equalization of temperature which has just taken place, the entropy will have again increased.

(3) Next, let the temperature be raised or lowered until the temperature of the initial state P is attained, the angular momentum and mechanical kinetic energy both being kept zero.

(4) Lastly, let the body be brought to its initial state by doing an amount of work upon it equal to T_P, the mechanical kinetic energy in the state P. This will not alter the entropy of the body.

Since the entropy has the same value at the end of a complete cycle as at the beginning, it follows that in the third operation the entropy must have decreased. In this operation, therefore, we have $\int \frac{dQ}{\theta}$ negative. Hence since it is à priori evident that the operation may be performed

in one or other of two ways, either so as to make every element of heat, dQ, positive, or so as to make every element negative, we conclude that the total amount of heat absorbed will be negative. Denoting it by $-Q$, where Q is a positive quantity, the principle of energy gives
$$T_p - T_q - Q = 0,$$
or
$$T_q - T_p = -Q,$$
from which we conclude that the mechanical kinetic energy continually decreases until the final invariable state is reached.

Let us consider the motion at any instant t, when G is the position of the centre of mass, Gz of the axis of figure, and Gz' of the axis about which the body is then rotating. Take Gx at right angles both to Gz and Gz'; and take

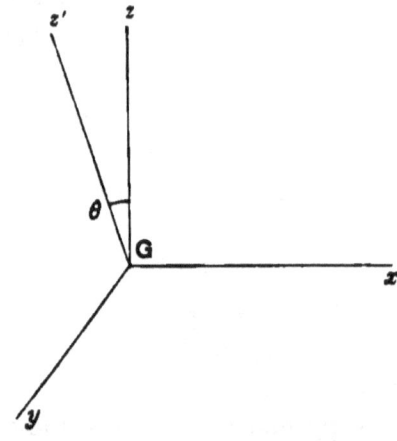

Gy, Gy' in the plane zGz', perpendicular respectively to Gz and Gz'. Then, if ω be the angular velocity of the body at the time t, and C' the moment of inertia about Gz', the angular momentum about Gz' will then be $C'\omega$; about Gy', $-\omega \Sigma mz'y'$; and about Gx, $-\omega \Sigma mz'x$, which is clearly zero. Thus the axis, GK, of resultant momentum at G, lies in the plane zGz', and the resultant momentum H is such that

$$H^2 = \omega^2 \{C'^2 + (\Sigma mz'y')^2\}.$$

Also the mechanical kinetic energy, T, at the time t, is given by

$$T = \tfrac{1}{2}MV^2 + \tfrac{1}{2}C'\omega^2,$$

where M is the mass of the body and V the velocity of its centre of mass.

Hence, since H and V are constant while T continually diminishes until the ultimate state is attained, we see that

$$\frac{C''^2 + (\Sigma mz'y')^2}{C'}$$

must continually increase up to the final state.

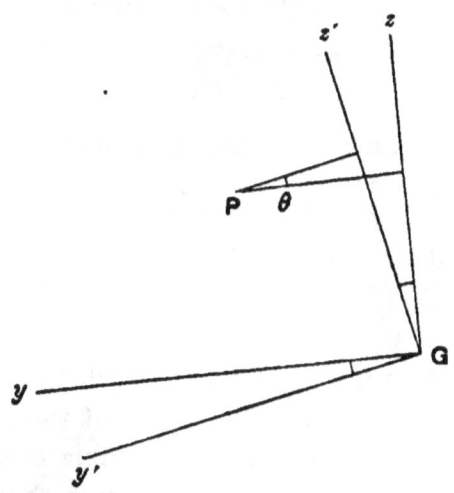

On account of the symmetrical form of the body, this result can be exhibited in a simpler form.

If we denote the angle between Oz and Oz', or the

equal angle between Oy and Oy', by θ, the coordinates of any point P will be connected by the relations

$$y' = y \cos \theta - z \sin \theta \\ z' = y \sin \theta + z \cos \theta \Big\}.$$

Thus, since Σmzy is clearly zero,

$$\begin{aligned} C' &= \Sigma m \, (x'^2 + y'^2) \\ &= \Sigma m \, (x^2 + y^2) + \sin^2 \theta . \Sigma m \, (z^2 - y^2) \\ &= \Sigma m \, (x^2 + y^2) + \sin^2 \theta . \Sigma m \, \{(z^2 + x^2) - (y^2 + x^2)\} \\ &= C + (A - C) \sin^2 \theta \\ &= C \cos^2 \theta + A \sin^2 \theta, \end{aligned}$$

where C is the moment of inertia about the axis of figure, and A that about any perpendicular axis.

Again, $\quad \Sigma mz'y' = \sin \theta \cos \theta . \Sigma m \, (y^2 - z^2)$
$$= (C - A) \sin \theta \cos \theta.$$

Hence $\quad C'^2 + (\Sigma mz'y')^2 = C^2 \cos^2 \theta + A^2 \sin^2 \theta$
$$= (A + C) C' - AC.$$

Thus we see that, until the ultimate state is reached, the value of $A + C - \dfrac{AC}{C'}$ must continually increase. Thence it follows that C' continually increases, and therefore, since $\omega^2 \{(A + C) C' - AC\}$ remains constant, that ω continually decreases.

Now since $C' = C + (A - C) \sin^2 \theta$, where A is evidently less than C, the moment of inertia about the axis of figure will be greater than about any other axis through G, and the moment of inertia about any axis through G will decrease as θ increases up to $\dfrac{\pi}{2}$. The effect of friction will therefore be to bring the axis of rotation nearer to the axis of figure; and when they once coincide, they will not again separate, so long as external influences are absent.

Furthermore, the instantaneous axis of rotation, Gz', can only coincide with the axis of resultant angular momentum at G when the angular momentum about any straight line through G at right angles to Gz' is zero. This can only take place when $\theta = 0$ or $\theta = \frac{\pi}{2}$, so that the axis of rotation must then either coincide with the axis of figure or lie in the equator.

Hence, since friction will prevent the axis of rotation from getting into the equator, we conclude that the effects of friction will not cease until the axis of rotation and the axis of figure coincide with the axis of resultant angular momentum at G.

If we were to observe the motion for a short time only, the effects of friction would be imperceptible, and C' and θ would be sensibly constant, so that the axis of rotation would appear to describe a cone in the body about the axis of figure. Consequently, if an ideal sphere be supposed carried about fixed in the body with its centre at G, the axis of rotation would intersect its surface in a circle. On watching the motion long enough, it would, however, be seen that the curve was not a perfect circle, but a nearly circular spiral gradually approaching the point in which the axis of figure meets the sphere.

To find the motion as it would appear for a short time, we must first find the position of GK in the plane zGz'.

Since when the angle zGz' is θ, the angular momentum about Gz' is $C'\omega$, and about Gy', $-\omega(C-A)\sin\theta\cos\theta$, the axis GK will lie on the side of Gz' remote from Gy', and make an angle ψ' with Gz', such that

$$\tan \psi' = \frac{(C-A)\sin\theta\cos\theta}{C'}$$

$$= \frac{(C-A)\sin\theta\cos\theta}{C\cos^2\theta + A\sin^2\theta}$$

$$= \frac{(C-A)\sin\theta\cos\theta}{(C-A)\cos^2\theta + A}$$

$$< \tan\theta.$$

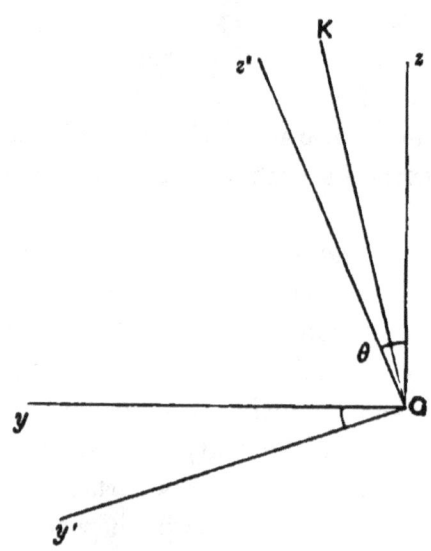

Thus GK lies between Gz' and Gz.

Again, if ψ be the angle between GK and Gz, we have

$$\psi = \theta - \psi',$$

and
$$\tan\psi = \frac{\tan\theta - \tan\psi'}{1 + \tan\theta \tan\psi'},$$

$$= \frac{A\tan\theta}{C}.$$

Hence also
$$\sin\psi = \frac{A\sin\theta}{\sqrt{C^2\cos^2\theta + A^2\sin^2\theta}},$$
$$\sin\psi' = \frac{\omega(C-A)\sin\theta\cos\theta}{H}$$
$$= \frac{(C-A)\sin\theta\cos\theta}{\sqrt{C^2\cos^2\theta + A^2\sin^2\theta}}.$$

Since we only require the motion relative to G, let us take G to be at rest; and suppose a fixed ideal sphere of unit radius described with its centre at G. Let GK, the axis of resultant angular momentum at G, meet the sphere in the fixed point K, and let Gz, Gz', the positions at the time t of the axes of figure and rotation, meet it in the points z, z'. Then if, at the consecutive instant $t+\tau$, the axes of figure and rotation meet the sphere in the points z_1, z'_1, these points will lie in a plane through GK, the arcs zz_1, $z'z'_1$, will be at right angles to the arc zKz', and

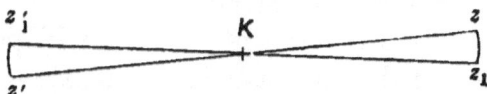

zz_1 will be equal to $\omega\tau\sin\theta$. The angle zKz' will therefore be equal to $\dfrac{\omega\tau\sin\theta}{\sin\psi}$; and the arc $z'z'_1$ to $\dfrac{\omega\tau\sin\theta}{\sin\psi}\sin\psi'$, or $\omega\tau\dfrac{C-A}{A}\sin\theta\cos\theta$. Hence the plane $zGz'K$ rotates in space about GK, in the same direction as ω, with angular velocity $\dfrac{\omega\sin\theta}{\sin\psi}$, or $\omega\dfrac{\sqrt{C^2\cos^2\theta + A^2\sin^2\theta}}{A}$. Also, since the same point of the body is situated at z' at the times t, $t+\tau$, the plane $zGz'K$ rotates in the body about the axis of figure, in the same direction as ω, with angular velocity

$$\omega \frac{C-A}{A} \cos \theta.$$

When θ is indefinitely small, these angular velocities become, respectively, $\dfrac{C\omega}{A}$ and $\dfrac{C-A}{A}\omega$, so that their ratios to ω are both finite.

If, secondly, we have a body so fluid that it can assume a form symmetrical with respect to every plane containing any given axis passing through its centre of mass, and if, after setting it in rotation in any way, we protect it from all external influences; it is evident that the parts of the body through which the axis of rotation will ultimately pass, will not be the same whatever may be the state of the body when first abandoned to itself. For, if the body be unelectrified and be held until it assumes a figure of revolution about the axis of rotation, it is clear that, in the invariable state assumed after the external influences have been removed, the axis of rotation will pass through the same parts of the body as when the body is first left to itself. Consequently, if we take any two such cases, in which the body is rotating about different axes when first abandoned to itself, the axis of rotation cannot pass through the same parts of the body in the two final states.

Again, if the body be allowed to assume an invariable state, and then be acted on for a time by external causes, the axis of rotation in the ultimate invariable state will not generally pass through the same parts of the body as before the disturbing forces came into operation. Hence, since the body will be symmetrical with respect to every plane through the axis of rotation in every invariable state assumed under the action of no external influences, the disturbing forces will generally have the effect of altering the relative positions of the parts of the body.

79. The surface of the earth is about three-fourths covered with a comparatively shallow, fluid ocean: the rest of the mass may be treated as 'rigid,' that is, as if the relative positions of its parts were invariable. The rigid nucleus appears to be practically symmetrical with respect to every plane through an axis Gz, where G is the centre of mass; while the ocean can change its place on the surface to suit the forces which act upon it. We therefore conclude that if the earth be protected from all external influences, or be merely allowed to radiate energy into space, it will be intermediate in behaviour between the two bodies considered in the last article. Thus, on account of the preponderating mass of the rigid nucleus, the ultimate axis of rotation will always be very nearly coincident with Gz, the axis of figure; and on account of the fluidity of the ocean, the rate at which the axis of rotation approaches Gz will be comparatively rapid.
In reality, the earth is acted on by the attractions of the sun, moon, and planets; and as these bodies are not situated in the plane of the earth's equator, their attractions will exert a small moment about an axis through G perpendicular to Gz. Thus if the whole of the earth were as rigid as the nucleus, the axis of rotation might, in time, separate considerably from the axis of figure; but the fluidity of the ocean will have a contrary tendency to keep the deviation of these axes small, or, as we may express it, as the axis of rotation moves about in space, it will be followed by the axis of figure.

13—2

The position of the sea with respect to the land is determined to a small extent by the inequalities of the surface of the nucleus, but chiefly by the centrifugal force due to the diurnal rotation, combined with the attraction of the rigid nucleus and with the attraction of the water on itself. The principal variable elements on which the form of the sea depends, are therefore the position of the axis of rotation with respect to the rigid nucleus, and the amount of the angular rotation. If the axis of rotation were inclined at any considerable angle to Gz, the axis of figure of the nucleus, these invariable elements would be undergoing continual changes, and in consequence, the ocean would be perpetually altering its distribution on the surface of the earth, in some places flooding the land, in others, leaving large tracts of sea dry. That such an event does not occur, evidently proves the ocean capable of keeping the two axes near together in spite of the efforts of the sun, moon, and planets, to make them separate.

At present, when the axis of rotation is practically coincident with Gz, the axis of figure, and the angular velocity equal to $\frac{2\pi}{24 \times 60^2}$, the surface of the earth is very approximately that generated by the revolution of an ellipse about its minor axis, the centre of the ellipse being placed at G, and the minor axis along Gz. The equation to this surface, in polar coordinates, is

$$r = a(1 - \epsilon \cos^2 \theta),$$

where G is the origin, Gz the initial line, and ϵ a constant which may be taken to be $\frac{1}{295}$.[1]

[1] Everett's 'Units,' or Herschell's 'Astronomy.'

CARNOT'S PRINCIPLE. 197

If the axis of rotation were not identical with the axis of figure, the form of the surface would be different. It is not within the province of thermodynamics to attempt an exact determination of the form that would be then assumed; but, merely to obtain some rough general ideas, we will suppose that when the angle between Gz', the axis of rotation, and Gz, the axis of figure, is small, and the angular velocity the same as before, the equation to the surface is still $r = a(1 - \epsilon \cos^2 \theta)$, but with Gz' for the initial line.

Take a plane section through Gz and Gz', and let any line through G in this plane, making an angle θ with Gz on the other side of Gz', meet the original surface in P and the new surface in Q. Then if α be the small angle zGz',

$$GP = a(1 - \epsilon \cos^2 \theta),$$
$$GQ = a\{1 - \epsilon \cos^2(\theta + \alpha)\}.$$

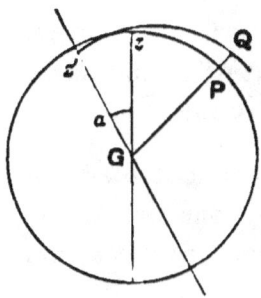

Hence the height the sea will rise at P on account of a small displacement of the axis of rotation from Gz, will be

$$PQ = a\epsilon \{\cos^2 \theta - \cos^2(\theta + \alpha)\}$$
$$= a\epsilon\alpha \sin 2\theta.$$

The distance zz' being $a\alpha$, we obtain
$$\frac{PQ}{zz'} = \frac{\sin 2\theta}{295}.$$
Thus, for a given value of zz', PQ will be a maximum when $\theta = \frac{\pi}{4}$, its value being then
$$\frac{zz'}{295}.$$
If zz' be a kilometre, the maximum value of PQ will be about 3·4 metres; if zz' be a mile, the maximum value of PQ will be about 18 feet.

From this calculation, it appears that a slight displacement of the axis of rotation from the axis of figure would cause a considerable change in the mean level of the sea in places in latitude 40° to 45°. This effect might be looked for in low-lying districts like the eastern parts of England or the country round the sea of Azof. No measurable oscillation in the mean level of the sea being known to exist, we are forced to conclude that the distance between the extremities of the axes of figure and rotation must always be very small.

80. We will next consider the influence of tidal friction in modifying the orbits of the heavenly bodies and the angular velocities with which they rotate about their axes. This subject appears to have been first noticed by Kant and Prof. J. Thomson[1], and more explicitly still, by Mayer. A few years after this (1854), it was pointed out by Helmholtz that the fact that the moon always turns the same face to the earth, is to be accounted for by the tides produced by the earth in the

---

[1] Tait's 'Sketch of Thermodynamics.'

moon while it was in a more fluid state than at present. Of late years, the effects of tidal friction have been studied by Prof. G. H. Darwin[1].

To make the discussion simple, let us imagine two separate bodies S, E, with no other body near them; and, to shorten the verbiage, let us suppose that their common centre of mass is at rest. Let S, E, be the centres of mass of the two bodies at any instant t; S', E', at a consecutive instant $t + \tau$. Then since SE, $S'E'$, pass through the common fixed centre of mass G, the four points (S, S', E, E') lie in one plane. Suppose, next, that the masses and mechanical motions of the two bodies are symmetrical with respect to this plane at the time t, and let the plane be taken to be that of the paper. Then clearly the centres of mass will continue to lie in the plane of the paper, and the masses and mechanical motions to be symmetrical with respect to it, so that we may represent the mechanical motions of each body at any instant with respect to its centre of mass by a rotation about an axis through that centre of mass at right angles to the plane of the paper, combined with certain irregular mechanical motions belonging to tides. Again, let the tides be at all times so small and the states of the bodies so stable that the axes of rotation may be considered fixed in the bodies and the moment of inertia of each body about its axis constant. Suppose also that, except for the tides, each body would be symmetrical with respect to every plane through its axis. Lastly, let the strata of equal density in both bodies be nearly spherical shells concentric with the centre of mass.

If, at any instant t, the tides of the body E be heaped

[1] Thomson and Tait's 'Natural Philosophy,' Vol. II.

up symmetrically about the two points in which the line SGE meets the surface of E, the attraction of S on E will then have no moment about the axis through E. Hence

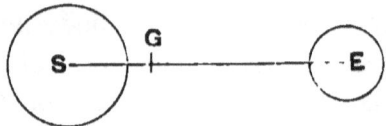

if we assume that the only forces which modify the angular momentum of the body E are due to the attraction of S, it will follow that, at the time t, the rate at which the angular momentum of E about its axis is increasing with the time, is zero. Consequently, if at the time t, E is moving as a rigid body with angular velocity of rotation ω_E, we shall have $\dfrac{d\omega_E}{dt} = 0$. Thus so long as E moves like a rigid body with its tides symmetrical about the line SGE, its angular velocity ω_E about its axis will remain constant; and therefore the orbital angular velocity of the point E about G, or, what is the same thing, of G about E, will be constant and equal to ω_E. The orbit of E about G will then evidently be a circle whose radius depends on ω_E.

If the orbit of E about G be not circular, the orbital angular velocity of E will not be constant: if the orbit be circular but not of the proper dimensions, the orbital angular velocity will be constant but not equal to ω_E. To illustrate what would happen in either of these cases, let us suppose that at the time t, the orbital angular velocity of E about G is less than ω_E, or equal to ω_E and decreasing. Then in a short time, the tides will have been carried in front of the line SGE; while, on the other hand, the attraction of S will tend to bring them back to that line. The attraction of S will therefore prevent them from fully

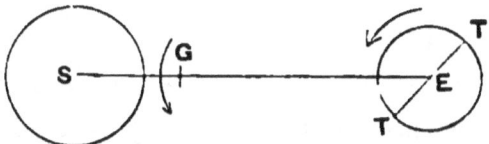

participating in the rotation of E about its axis, so that they will act as a friction brake on the motion. Similarly, if the orbital angular velocity at the time t be greater than ω_E, or equal to ω_E and increasing, the tides will soon be left behind the line SE, while the attraction of S will tend to drag them on. Thus they will again have a frictional action, but this time they will accelerate the rotation of E on its axis.

It will thus be seen that the ultimate effect of the tides on the system will be to make it move as rigid, whether the two bodies are ultimately fused into one, or are merely in contact, or whether they continue to move as distinct bodies at a distance from one another under the influence of gravitation alone. In any case, all parts of the system will ultimately describe circular orbits about G with the same constant angular velocity.

Suppose now that, at the time t, the two bodies are rotating about their axes without mechanical vibrations,

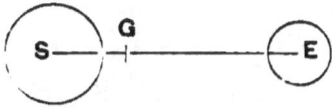

with angular velocities ω_S, ω_E, respectively; and the points S, E, about G with common angular velocity ω. Also let the masses of the two bodies be (S, E); their moments of inertia about their axes through S and E, (C_s, C_E); and denote the distances GS, GE, by R and r. Then the angular momentum H, of the system about the axis through G, perpendicular to the paper, is

$$H = (SR^2 + Er^2)\omega + C_s\omega_s + C_E\omega_E.$$

Again, if we assume that the only parts of the energy of the system that can vary, are the mutual potential energy of the two bodies, and the mechanical and non-mechanical kinetic energies, it can easily be proved, by Carnot's principle, that so long as frictional actions exist, the sum of the mutual potential energy of the two bodies, and of the mechanical kinetic energies, must continually diminish. In popular language, this would be expressed by saying that the mechanical kinetic energies and mutual potential energy are partially converted, by tidal friction, into the non-mechanical kinetic energy of 'heat.'

To facilitate calculation, let us suppose that at the time t, the points S, E are describing circular orbits about the point G. Then since the mutual attraction between S and E is approximately the same as if they were concentrated into particles at their respective centres of mass, we have, nearly,

$$\left. \begin{array}{c} \dfrac{\lambda E}{(R+r)^2} = R\omega^2 \\ \dfrac{\lambda S}{(R+r)^2} = r\omega^2 \end{array} \right\} \quad \ldots\ldots\ldots\ldots\ldots(47).$$

But since $\qquad SR = Er = x$ (say),

we find $\qquad R + r = \dfrac{S+E}{E} R = \dfrac{S+E}{S} r$,

and $\qquad x = \dfrac{SE}{S+E}(R+r).$

Hence
$$\omega^2 = \frac{\lambda(S+E)}{(R+r)^3} \quad \ldots\ldots\ldots\ldots\ldots(48),$$

and
$$(SR^2 + Er^2)\omega = \sqrt{\frac{\lambda S^2 E^2}{S+E}(R+r)^{\frac{1}{2}}} \quad \ldots\ldots\ldots(49).$$

Again, the sum of the mechanical kinetic energies of translation at the time t is

$$\tfrac{1}{2}(SR^2 + Er^2)\omega^2 = \tfrac{1}{2}\frac{\lambda SE}{R+r} \quad \ldots\ldots\ldots\ldots\ldots(50);$$

and lastly, the mutual potential energy of the two bodies is

$$\int_{R+r}^{R_0+r_0} \lambda \frac{SE}{r^2}(-dr) = \lambda SE \left(\frac{1}{R_0+r_0} - \frac{1}{R+r}\right) \ldots\ldots(51),$$

where R_0 and r_0 are the values of R and r in the standard state.

If the two bodies preserve their properties, and are not in contact in the final state, so that they are compelled to describe their orbits about G by the force of gravity alone, it is obvious that the formulae just obtained will apply to the final state as well as at the time t.

Now let us make the supposition that at the time t, the points E and S are rotating about G in the positive direction, or that ω is positive. This will not detract from the generality of our conclusions. Furthermore, let us denote the final value of ω by Ω, and take the three following illustrative cases of the ultimate effects of the tides.

(1) Let H, and consequently Ω, be positive, and suppose both ω_s and ω_E greater than Ω. Then tidal friction will diminish both ω_s and ω_E, so that $(SR^2 + Er^2)\omega$ will increase. Hence, by equation (49), $R + r$ will increase, and therefore, by equation (48), ω will decrease, or the periodic time be lengthened. Again, we see that the mechanical kinetic energies of translation and rotation

decrease, the diminution being accounted for, partly by the frictional generation of non-mechanical kinetic energy, and partly by the increase of the mutual potential energy of the two bodies.

(2) Let H and Ω both be positive, and both ω_s and ω_E less than Ω. In this case, tidal friction will increase both ω_s and ω_E; and consequently diminish $R + r$, or draw the bodies nearer together and shorten the periodic time. The mechanical kinetic energies of translation and rotation will both increase; and the non-mechanical kinetic energy generated by tidal friction will be produced entirely at the expense of the mutual potential energy of the two bodies.

(3) If H be negative, Ω will also be negative, or the two centres of mass will ultimately describe their orbits about G in the negative direction. This result would be brought about in the following manner. The two bodies would be drawn together by tidal friction until they actually came into contact. If the directions of the orbital motions were not reversed when the bodies separated, they would again be drawn together; and so on.

It might, however, happen that the two bodies are partially or wholly fused, or even converted into vapour, by the violence of the grinding or impulsive actions which take place when they come together. But whether they are fused into one, or preserve their identities, or are broken up into several parts, the whole system will ultimately rotate about the axis through G like one rigid body.

81. In the solar system, almost the whole of the angular momentum is, at present, in one direction. Hence if we neglect the attraction of the fixed stars, it follows that, when the motions are ultimately reduced to regularity, the whole system will rotate together in this direction.

To obtain a rough estimate of the mutual actions of the sun and earth, let us consider the two bodies S, E, of Art. 80, calling S the sun and E the earth; and, for the sake of simplicity, let us imagine all other bodies to be absent. Then tidal friction will tend to make the orbits of S and E perfectly circular, and also to cause the two bodies to turn the same faces to one another. These effects will take place simultaneously, but for our purpose we may suppose them to take place separately; the orbits becoming circular before there are any perceptible changes in the rotations.

Since $\dfrac{SR^2}{Er^2} = \dfrac{R}{r} = \dfrac{E}{S}$, which is exceedingly small, we may neglect the angular momentum of translation of the sun in comparison with that of the earth. We may therefore treat S, the centre of mass of the sun, as a fixed point, and E, the centre of mass of the earth, as a moving point. Hence so long as the rotations are supposed unaffected, or the angular momentum of translation of the body E about S constant, we may apply Kepler's laws. Thus if (a, b) be the semi-axes of the ellipse described by E about S, e its eccentricity, and T the periodic time, or length of the year, Kepler's third law gives

$$T^2 \propto a^3.$$

But if $\dfrac{dA}{dt}$ be the rate of description of areas about S by the line SE,

$$T = \dfrac{\pi ab}{\dfrac{dA}{dt}}.$$

Hence
$$\left(\frac{dA}{dt}\right)^2 = \frac{\pi^2 a^2 b^2}{T^2}$$
$$\propto \frac{b^2}{a}$$
$$\propto a(1-e^2).$$

Now if v be the velocity of the point E about S, ds an element of the orbit, and p the perpendicular from S on the tangent, we have evidently
$$\tfrac{1}{2}p\,ds = dA,$$
and
$$\tfrac{1}{2}pv = \frac{dA}{dt}.$$

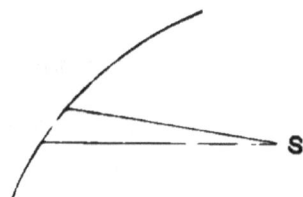

But Epv is the angular momentum of translation of the body E about the fixed point S, which is supposed constant. Hence the value of $a(1-e^2)$ remains constant; and therefore, if a' be the radius of the orbit when it has become circular,
$$a' = a(1-e^2).$$
Taking the present value of e to be $\frac{1}{60}$, we see that when the orbit becomes circular, the mean distance will have diminished by the 3600th part, and the length of the year by the 2400th part, that is, by 3·65 hours.

While the orbit of E about S is thus becoming circular, the angular velocity of translation will never vary much from its present value. Hence when the circular motion commences, the angular velocity of translation will be considerably less than both those of rotation. Tidal friction will then evidently check both of the axial rotations, drive the earth further away from the sun, and increase the length of the year. The consequent diminutions in the angular momenta of rotation will be attended by an exactly equal increase in the sum of the angular momenta of translation; and the mechanical kinetic energies of rotation and translation will decrease, partly to supply the increase in the mutual potential energy of the two bodies, and partly to supply the non-mechanical kinetic energy developed by tidal friction.

In making a rough comparison of the effects of tidal friction on the angular momenta of rotation of the sun and earth, it will be sufficient, on account of the great distance between the two bodies, to treat them as spheres of radii a_s, a_e.

The tides are due to the fact that neither body attracts all parts of the other alike. Thus the attractions of the earth on masses of one gramme, situated on the nearest and remotest parts of the sun's surface, are respectively $\dfrac{\lambda E}{(R+r-a_s)^2}$ and $\dfrac{\lambda E}{(R+r+a_s)^2}$, the difference between which is roughly $\dfrac{2\lambda E}{(R+r)^2} \cdot \dfrac{a_s}{R+r}$. Hence since each of the masses is attracted by the sun with a force $\dfrac{\lambda S}{a_s^2}$, we may take $\dfrac{2E}{S} \left(\dfrac{a_s}{R+r} \right)^3$ as a measure of the tide-producing force on the sun. Similarly, $\dfrac{2S}{E} \left(\dfrac{a_e}{R+r} \right)^3$ will represent the tide-producing force on the earth. Consequently, the tide-producing force on the earth exceeds that on the sun in the ratio $\left(\dfrac{S}{E}\right)^2 \left(\dfrac{a_e}{a_s}\right)^3$. Denoting the sun's mean density by ρ_s, and the earth's by ρ_e, this ratio is equal to $\dfrac{S}{E}\dfrac{\rho_s}{\rho_e}$, or about 94000.[1] We therefore conclude that the tides will be much higher on the earth than on the sun.

Again, not only are the tides much smaller on the sun than on the earth, but the force of gravity is greater on the surface of the sun than on that of the earth in the ratio $\dfrac{S}{E}\left(\dfrac{a_e}{a_s}\right)^2$, or $\dfrac{a_s \rho_s}{a_e \rho_e}$, or 29. It may therefore be inferred that if the sun and earth be removed to a considerable distance from one another without allowing the tides to undergo any sensible alterations, and the two bodies be then set at liberty, the tides would subside much more rapidly on the sun than on the earth. Thence it may be assumed that when the two bodies are rotating freely with equal angular velocities within the range of one another's influence, the tides that will be produced will be much more nearly symmetrical about the line SE in the case of the sun than in the case of the earth.

Thus so long as the earth revolves on its axis with any moderate speed, tidal friction will have a much more energetic effect on the angular momentum of the earth's rotation than on that of the sun.

In general, if two unequal bodies of the same liquidity and density are rotating about their axes with the same angular velocity, and be placed in presence of each other, the rotation of the smaller body will be modified by tidal friction much faster than that of the other, because the

---

[1] The values of ρ_s, ρ_e are calculated from the data given in Herschel's 'Astronomy.'

angular momentum of rotation is less, and the force which tends to modify it greater, than in the case of the larger body. This is probably the reason why the moon has been already caused to turn the same face to the earth while the earth does not yet turn the same face to the moon.

Returning to the case of the sun and earth, we see that, after the orbits become circular, the increase in the sum of the angular momenta of translation will, at first, be chiefly due to the decrease of the diurnal rotation of the earth. This will go on until the earth always presents the same face to the sun, and then it will proceed, at a much slower rate, at the expense of the rotation of the sun.

It will, of course, be easy to calculate what will be the distance between the earth and sun when they show the same faces to one another; but we may also estimate the distance when the earth first shows the same face to the sun, by assuming that, until that time, tidal friction produces no effect on the rotation of the sun. For if we take the earth's 'radius of gyration' to be a third of her radius, and remember that the ratio of r to $R+r$ is always very nearly unity, we have roughly, when the orbits first become circular,

$$\frac{C_E}{SR^2 + Er^2} = \frac{C_E}{Er^2}$$

$$= \left(\frac{1}{70000}\right)^2$$

$$= \frac{1}{49 \times 10^9}.$$

Hence the ratio of the angular momentum of the earth to

the sum of the angular momenta of translation of the earth and sun, will at first be about

$$\frac{365}{49 \times 10^8},$$

or

$$\frac{1}{13 \times 10^6}.$$

Now since the orbital angular velocity of the earth about the sun decreases as the circular orbit increases and is already very small, it is clear that when the earth is caused to turn the same face to the sun, she will have practically lost the whole of her angular momentum of rotation. Hence from the time that the orbits first become circular to the time when the earth begins to turn the same face to the sun, the sum of the angular momenta of translation will have increased by its $\frac{1}{13 \times 10^6}$ th part, and therefore, by equation (49), the distance between the two bodies will have increased by its $\frac{2}{13 + 10^6}$ th part.

Thus the sum of the angular momenta of translation and the distance between the earth and sun, will, on our suppositions, be scarcely changed until tidal friction begins to affect the rotation of the sun.

Again, if we take the sun's radius of gyration to be a third of his radius, we have, at present,

$$\frac{C_s}{SR^2 + Er^2} = \frac{C_s}{Er^2} = \frac{S}{E}\left(\frac{1}{630}\right)^2,$$

which is rather less than 1.

Thus the ratio of the sun's angular momentum of rotation to the sum of the angular momenta of translation, or $\frac{C_s \omega_s}{(SR^2 + Er^2)\omega}$, is at present about $\frac{365}{25}$, or 14·6.

CARNOT'S PRINCIPLE. 211

Hence when the earth and sun show the same faces to one another, the sum of the angular momenta of translation will be 15·6 times as large, and their distance 243 times as large, as they are now. The year will then be 3796 times as long as at present.

If the planet E had been mercury instead of the earth, the last result would have been quite different. For we should then have had, at present,

$$\frac{C_s}{Er^2} = \frac{S}{E}\left(\frac{1}{240}\right)^2$$

$$= 84, \text{ nearly.}$$

Consequently the ratio of the sun's angular momentum of rotation to the sum of the angular momenta of translation of the sun and mercury, is at present

$$\frac{84 \times 88}{25},$$

or about 295.

Hence when the two bodies are caused, by tidal friction, to rotate as one, the distance between them, on the supposition that all other bodies are absent, will be 87000 times as great as it is now.

In comparing the changes produced by tidal friction in the mechanical kinetic energies, after the orbits have been rendered circular and before E turns the same face to S, let us denote corresponding small increments by the letter d. Then $d(\frac{1}{2}C_E\omega_E^2) = C_E\omega_E d\omega_E = \omega_E d(C_E\omega_E)$, and $d(\frac{1}{2}C_s\omega_s^2) = \omega_s d(C_s\omega_s)$. Thus

$$\frac{d(\frac{1}{2}C_s\omega_s^2)}{d(\frac{1}{2}C_E\omega_E^2)} = \frac{\omega_s}{\omega_E}\frac{d(C_s\omega_s)}{d(C_E\omega_E)}.$$

Now $\frac{\omega_s}{\omega_E}$ will increase, but even if we suppose ω_s to remain

14—2

unchanged until the earth turns the same face to the sun, its value cannot exceed $\frac{365}{25}$, or 14·6. On the other hand, so long as the earth possesses any considerable angular rotation, the value of $\frac{d(C_s\omega_s)}{d(C_E\omega_E)}$ will be exceedingly small. Hence the effects of tidal friction on the mechanical kinetic energy of rotation will be negligible in the case of the sun compared with that of the earth.

Again, if U_t be the sum of the mechanical kinetic energies of translation and H_t the sum of the angular momenta of translation, equations (49) and (50) give

$$dU_t = -\tfrac{1}{2}\frac{\lambda ES}{(R+r)^2} d(R+r),$$

$$dH_t = \tfrac{1}{2}\sqrt{\frac{\lambda E^2 S^2}{S+E}}\frac{d(R+r)}{(R+r)^{\frac{1}{2}}},$$

so that

$$dU_t = -\frac{\sqrt{\lambda(S+E)}}{(R+r)^{\frac{3}{2}}} dH_t$$

$$= -\omega dH_t,$$

a negative quantity.

But so long as we can neglect the changes in the sun's angular momentum of rotation,

$$C_E\omega_E + H_t = \text{constant},$$

or

$$d(C_E\omega_E) + dH_t = 0.$$

Hence

$$dU_t = \omega d(C_E\omega_E)$$

$$= \frac{\omega}{\omega_E} d(\tfrac{1}{2} C_E\omega_E^2).$$

Now when the circular orbit undergoes a small increase, the increase in the mutual potential energy of the two bodies is $\frac{\lambda ES}{(R+r)^2} d(R+r)$, and the decrease in the sum of the mechanical kinetic energies just half as much. We therefore see that the whole of the decrease in the sum of the mechanical kinetic energies of translation, and the fraction $\dfrac{\omega}{\omega_E}$ of the decrease in the mechanical kinetic energy of the earth's rotation, will be required to supply the increase in the mutual potential energy, and that only the remaining part, $1 - \dfrac{\omega}{\omega_E}$, of the mechanical kinetic energy of rotation lost by the earth will be converted into non-mechanical kinetic energy by the friction of the tides. Since $\dfrac{\omega}{\omega_E}$ will have its least, and $1 - \dfrac{\omega}{\omega_E}$ its greatest value, just after the orbits become circular, it follows that the non-mechanical kinetic energy developed by tidal friction can never exceed the $\dfrac{364}{365}$th of the mechanical kinetic energy lost by the earth's rotation.

82. The effect of tidal friction in retarding the rotation of the earth about its axis, that is, in increasing the length of the day, is too minute to be detected by instrumental means; but it has been shown to exist by the calculations of Adams and Delaunay, which could not be made to agree with observation when the length of the day was supposed strictly constant. In the case of the other planets, their periods of rotation about their axes are, at present, only roughly known. We cannot therefore expect to have any independent evidence of the effects of the tides on their speeds of rotation.

Tidal friction affects the length of the year, partly by making the orbit circular, and partly by increasing its dimensions. From the former cause, the year may be shortened by 3·65 hours; from the latter it will be lengthened, but the increase will be small even by the time the earth is caused to turn the same face to the sun. Upon the whole we conclude that, if no other body were near the sun and earth, the length of the year, owing to tidal friction, would be slowly shortening.

23. As a nearer approximation to the actual state of the solar system, let us suppose that there are two planets E J revolving about S. Then if we can neglect the mutual attraction of E and J in comparison with the forces with which they are attracted to S, it is clear that if E and J be describing circular orbits of different sizes,

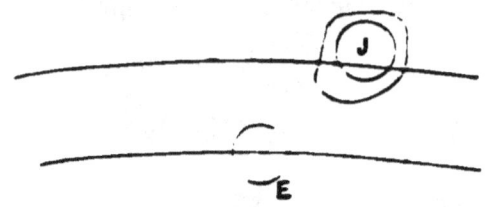

their periods of revolution will be different. It will therefore be impossible for the system of three bodies to be moving as rigid, and the motion will consequently be attended by tidal friction. In this case, we have not merely the mutual tidal influence of S and E and of S and J but of E and J as well. The effects produced by

It is shown in the planetary theory that, on the supposition that all the heavenly bodies are perfectly rigid, the disturbing effects of the other planets would cause the eccentricity of the earth's orbit to vary periodically, increasing to a maximum, then decreasing, then again increasing, and so on. The actual change in the length of the year is thus due to the joint agency of the disturbing actions of the other heavenly bodies and the slow effects of tidal friction.

the last-mentioned tides in any moderate length of time will, of course, be quite insignificant; but if sufficient time be allowed, they may accumulate to a considerable amount. We may easily obtain a rough idea of these effects, since we may suppose that the two planets have no tidal influence on one another except when they are comparatively near together, and may then represent the tides produced in one planet, say J, by the other planet E, as a deformation from the natural spherical shape to a slightly prolate, or elongated, spheroid with its axis pointing roughly to E, as shown on an exaggerated scale in the figure.

As similar tides will be produced in E by J, it is evident that the mutual attraction between E and J will be slightly greater than if the tides were completely absent. The effect of the mutual tidal influence of E and J will thus be to draw them nearer together, or to diminish the orbit of the outer planet and increase that of the inner.

When the whole solar system has been reduced, by tidal friction, to move as one rigid body, let any number of planets E, J,......be revolving round the sun in such a way that the force with which any one of them is attracted to the sun is incomparably greater than the other forces which act upon it. Then if ω be the common angular velocity of the system, and r_e, r_j, ...the radii of the circular orbits, we have

$$\frac{S}{r_e^2} = r_e\omega, \text{ or } S = r_e^3\omega\,;$$

$$S = r_j^3\omega\,;$$

$$\ldots\ldots$$

Hence $\qquad r_e = r_j = \ldots\ldots$

216   ELEMENTARY THERMODYNAMICS.

If two planets E', J', be so near together that, in the case of one or both of them, their mutual attraction cannot be neglected in comparison with the attraction of the sun, it will be evident from the figures, since the

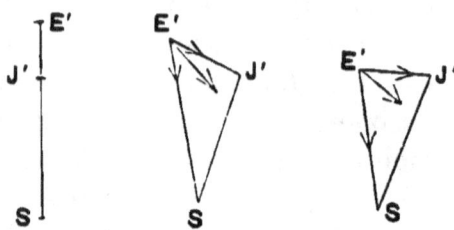

resultant force on every isolated planet must pass through the centre of mass of the whole system, which is practically the same as the centre of S, that the two planets E', J', must lie in a straight line with S.

It will now be clear that the planets will ultimately form a ring round the sun, since those planets which are ultimately at a considerable distance from one another, will be at the same distance from the sun, and since those planets which are behind one another will be comparatively near together. This ring will resemble, in some respects, those which now surround the planet Saturn; the chief difference being that, whereas the number of small bodies of which Saturn's ring is supposed to be formed, must be very great, the planets which will constitute the ring round the sun will be but few.

It is interesting to notice that we can conceive of no ultimate arrangement of the planets which would admit of these small bodies continuing to form circular rings around Saturn, as at present. We therefore conclude that the beautiful rings of Saturn will be broken up. For a long time to come, however, their existence will probably be preserved, in spite of the adverse influences of the rest of the solar system, by the friction of the tides due to the rapid axial rotation of Saturn.

The size and period of rotation of the ring which will ultimately encircle the sun can be readily calculated. For it can easily be shown that, at present, the sun contributes less than 2 per cent. of the total angular momentum of the solar system about its centre of mass. Hence since the rotation of the sun about his axis in the final state of the system will be vastly smaller than now, we see that the ultimate planetary ring will contain practically the whole of the angular momentum. Again, it may be shown that, at present, Jupiter, by his revolution round the sun, contributes 60 per cent. of the total angular momentum of the system, while his mass is 74 per cent. of the sum of all the planetary masses. If therefore the ultimate planetary ring were coincident with the present orbit of Jupiter, the total angular momentum of the solar system would only be about 80 per cent. of what it is now. The radius of the ring will therefore be about $(\frac{5}{4})^2$ times Jupiter's present distance from the sun, that is, 1200 million kilometres, or 750 million miles; and its period of rotation will be about $(\frac{5}{4})^3$ times the present length of Jupiter's year, that is, nearly 22 of our present years.

We can now predict the fate of the moon, on the assumption that no other planetary bodies ever come so near the earth and moon as to exert a sensible disturbing force on them. For if S be the mass of the sun, M that of the moon, r_s the present distance of the earth from the sun, and r_m its present distance from the moon, the tide-producing force on the earth due to the action of the sun, is, at present, to the tide-producing force due to the action of the moon, in the ratio $\dfrac{S}{M}\left(\dfrac{r_m}{r_s}\right)^3$, which is about $\dfrac{9}{16}$. Hence it is clear that the angular momentum lost by the earth's rotation goes partly to enlarge the orbit of the earth about the sun, and partly to enlarge the orbit of the moon about the earth. Even if the whole of the angular momentum of rotation lost by the earth went to drive the moon further away, the effect would not be considerable. For if E be the mass of the earth; $k\,(\equiv \tfrac{1}{3}$ the radius) the radius of gyration; ω_E the present angular velocity of the earth's axial rotation; and ω_m the present angular velocity of the moon's revolution about the earth; the ratio of the angular momentum of rotation of the earth to the moon's revolutional angular momentum about the earth, is, at present,

$$\dfrac{Ek^2\omega_E}{Mr_m^2\omega_m} \equiv 88 \times \left(\dfrac{1}{180.8}\right)^2 \times \dfrac{82}{3}$$

$$\equiv \dfrac{8}{109}.$$

Thus even on the supposition that the sum of the angular momenta of the earth and moon about their common centre of mass remains constant, the distance of the moon from the earth can never exceed its present value by more than the $\tfrac{16}{109}$th part; and the length of the month can not become more than $1\tfrac{2}{9}$ times as great as at present, that is, can not exceed $33\tfrac{1}{2}$ of our present mean solar days.

Since if the earth and moon continue separate, the month and the year must ultimately become of the same length, and therefore equal to 21 or 22 of our present years, it is clear that the moon must fall into the earth. This may also be shown as follows. Suppose, if possible, that the sun, earth, and moon, are in a straight line with one another and moving like a single rigid body. Then if the distance of the moon from the sun be $1 + x$ times the distance of the earth from the sun, where x is either positive or negative, the acceleration of the moon will be $1 + x$ times the acceleration of the earth. But if x be a small quantity and P the attraction of the sun on each gramme of the earth, the attraction of the sun on each gramme of the moon will be $\dfrac{P}{(1 + x)^2}$, or $P(1 - 2x)$. Hence, since we may neglect the attraction of the moon on each gramme of the earth in comparison with the attraction of the earth on each gramme of the moon, the attraction of the earth on each gramme of the moon must be $\pm 3Px$, or the earth's mass $\pm 3x^3$ times that of the sun. Thus

$$\pm 3x^3 = \frac{1}{359550},$$

or
$$x = \pm \frac{1}{328}.$$

Since the 328th part of the ultimate distance of the earth from the sun is several times greater than the greatest possible distance of the moon from the earth, we conclude, as before, that the moon cannot remain permanently separate from the earth.

It is easy to see how the moon and earth will be brought together. For a long time to come, the tides will have the effect of driving the moon further away from the earth. This will go on until the day is nearly as long as the month, and then the lunar tides will practically cease and the distance of the moon become stationary. Owing to the continued action of the solar tides, the day will at length become so much longer than the month that lunar tides will again be called into existence and the moon begin to approach the earth. The action of the tides will then transfer angular momentum from the moon's revolution about the earth to the earth's axial rotation, and from the latter to the earth's orbit about the sun. As the moon's orbit about the earth contracts, the two bodies will begin to deviate from moving as rigid, and they will at last come into violent collision. This conclusion will be made clear by the following argument. As the two bodies approach one another, let them be supposed to move as one rigid body. Then if H be the sum of their angular momenta about their common centre of mass, and r the distance between them, we have, by equations (48) and (49),

$$H = \sqrt{\frac{\lambda E^2 M^2}{E+M}} r^{\frac{1}{2}} + E\sqrt{\lambda}(E+M)\frac{k^2}{r^{\frac{3}{2}}},$$

or, nearly, $\quad H = M\sqrt{\lambda E} r^{\frac{1}{2}} + \dfrac{E\sqrt{\lambda E}k^2}{r^{\frac{3}{2}}},$

where the roots are to be taken positively.

Hence $\quad dH = \tfrac{1}{2} M \sqrt{\dfrac{\lambda E}{r}} \left\{ 1 - 3\dfrac{E}{M}\left(\dfrac{k}{r}\right)^2 \right\} dr$

$\quad\quad\quad = \tfrac{1}{2} M \sqrt{\dfrac{\lambda E}{r}} \left\{ 1 - 264 \left(\dfrac{k}{r}\right)^2 \right\} dr.$

Thus if dH and dr are both to be negative, r will have to be greater than $k\sqrt{264}$, that is, greater than $16.2k$. Consequently if we take k to be $\tfrac{1}{3}$ of the earth's radius, the least possible value of r will be 5·4 times the earth's radius, that is, 34,500 kilometres, or 21,400 miles. When this value of r is reached, the motion which we have assumed for the system ceases to be possible.

The only other planet we shall here consider is Uranus, whose satellites are the only bodies in the solar system with the exception of comets which are known to possess a retrograde motion. In this case, by supposing the tides produced by the satellites and by the sun to act alternately for equal short times, it will easily be seen that the satellites will be continually drawn nearer the primary, whose axial rotation, though it will be checked, cannot be rendered retrograde. Finally, the satellites will all fall into the primary and the whole motion will become direct.

84. Another effect of friction worthy of notice is that due to the action of running water and glaciers; but between these agencies and tidal friction there is an important difference, that, whereas the tides are merely due to want of 'perfect rigidity,' and may exist in any body, whether solid, liquid, or gaseous, the effects of running water and glaciers will rapidly diminish and disappear when, owing to the cooling down of the sun, the seas become frozen up, and rain and snow cease to fall.

Running water and glaciers are continually at work wearing away the land and transporting the detritus to lower places or to the sea, where fresh land is gradually forming. If this action should last long enough, without anything to counteract it, the land would, in time, be everywhere reduced to the level of the sea. However, there are two agents in operation which have a modifying influence. In the first place, the earth's crust is liable to be upheaved by volcanic and similar forces; and secondly, the slackening of the earth's rotation about its axis, owing to tidal friction, enables the waters of the sea, so long as they remain unfrozen, to slowly flow from the equator towards the poles, and so tends to submerge the land about the poles, and to make it more elevated in the neighbourhood of the equator. Now on referring to works on geology, it will be seen that the denuding effects of running water are amply sufficient to neutralize the tendency of tidal friction to cause an equatorial elevation of the land. We therefore see that, if it were not for volcanic and similar agencies, the earth would acquire a more even surface, with an increased quantity of land in the equatorial regions; and that, in the polar regions, the land would, in some places, be submerged by the rising of the sea, and in others, new land would be formed by the deposits of rivers and glaciers.

85. The effect of tidal friction in shortening the periodic time by making the orbit more circular, may be

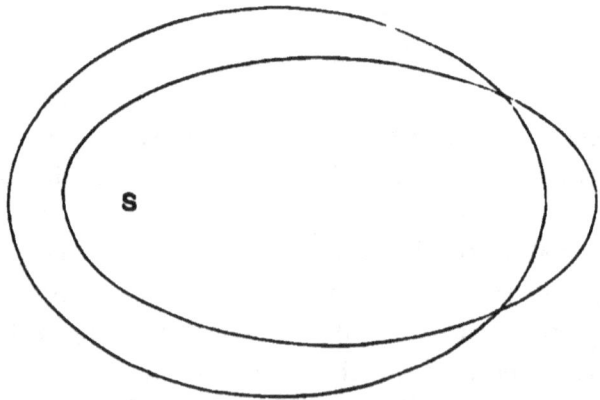

expected to be most considerable in the case of very elongated elliptical orbits, like those of comets. For, since by Kepler's second law, the line joining the comet to the sun describes equal areas in equal times, it is obvious that the comet's orbital angular velocity will be so extremely small in the more distant parts of the orbit that almost the whole of the periodic time will be spent in traversing these parts. Hence tidal friction, by merely making the orbit slightly less elongated, and consequently diminishing those parts of it which require the most time for their description, may produce a sensible diminution in the periodic time.

A change in the periodic time has only yet been observed in the case of Encke's comet, and the observed fact is a shortening of the period. This may be due, partly to tidal friction, and partly to solar radiation, as explained in the foot-note[1]; but a different explanation, propounded by Encke himself, has hitherto been almost universally accepted, at least, until quite recently. It was supposed that the ether offered a slight resistance to the mechanical motion of the comet through space, independent of the non-mechanical motions (see Art. 59), from which it would follow that those parts of the orbit which require most time for their description, would be cut shorter every revolution, just as by tidal friction; and in consequence, that the periodic time would diminish.

---

[1] When a comet approaches the sun, it is generally observed to throw off a tail, which points away from the sun, and is not merely left behind by the comet. The cause of the phenomenon is evidently equivalent to a repulsive force residing in the sun. Such a force will have no effect on the angular momenta of the comet; but, since the tail only appears when the comet is in that part of its orbit nearest the sun, it will make the orbit more circular, just like tidal friction, but perhaps more energetically.

We have already stated how we consider it probable that the tails of comets are produced by solar radiation.

The mere fact that the periodic time of a single comet is undergoing a slow diminution, appears to be a slender foundation for Encke's hypothesis of an etherial resistance to mechanical motion, and accordingly, there has lately been a disposition to reject it[1]. Now that other causes have been shown capable of explaining the comet's peculiar behaviour, we must admit that the evidence in favour of Encke's theory is very slight indeed[2]; and for this reason, it has not hitherto been taken account of in this work. Still, as it has not been completely disproved, it is necessary to allow the possibility of it, and to examine what would be its effects on the state of the solar system. This question is interesting in itself and historically important, partly from the frequent references that are made to it, and also because Encke's theory was universally accepted at the time that the final state of the solar system was predicted by Sir W. Thomson.

If we assume that the only external influences to which the solar system is subject are the resistance of the ether and the radiation of energy into space, then, in the final state, every part will be at the temperature of absolute zero, there will be no mechanical motions, and the different bodies will all lie together, instead of

---

[1] Newcomb's 'Popular Astronomy.'

[2] A theoretical argument which is sometimes thought to support Encke's hypothesis, is given in the late Prof. Balfour Stewart's 'Treatise on Heat.' By assuming that the laws of radiation are exactly the same for bodies in mechanical motion as for bodies in a state of mechanical rest, it is concluded that the principle of energy and Carnot's principle cannot both be true unless there be an etherial resistance to mechanical motion. The obvious reply to this argument is that the laws of radiation cannot be quite the same for bodies in rapid mechanical motion as for bodies in a state of mechanical rest.

being separated by considerable distances, as at present. Consequently, in the final state, there will be neither mechanical nor non-mechanical, kinetic energy of matter; but, owing to the force of gravitation, there will be potential energy, that is, ethereal kinetic energy bound to the material system.

Again, if we take any number of spherical bodies, the mass M, of any one of them whose radius is r and density ρ, will be $\frac{4}{3}\pi r^3 \rho$; while the area A, which will be exposed to the effects of a resisting medium, will vary as r^2. We therefore have $A = k\left(\frac{M}{\rho}\right)^{\frac{2}{3}}$, where k has the same value for all the bodies. Hence, if we make the usual assumption that, for a given velocity of the centre of mass, the resistance $\propto A$, it is clear that the time required by the resistance to diminish a given velocity by a given amount, will $\propto (M\rho^2)^{\frac{1}{3}}$.

Now the sun and planets immensely exceed a comet in mass, and are probably not inferior in density. The resistance of the ether will therefore have a vastly more important influence on the motion of a comet than on that of the sun or a planet. But, according to Encke's theory, the effect of the resistance is small, even in the case of a comet. Hence in the case of the sun and larger planets, it will be so insignificant that it may be supposed not to come into operation so long as there is any appreciable amount of tidal friction at work, and at all times the mechanical motions relative to the centre of mass will be sensibly the same as if that point were at rest.

As soon as the whole system begins to move practically as a rigid body, the resistance of the ether will come into action and cause the different bodies to describe nearly circular spirals. The motion will continue to be of this character until a collision occurs, or one of the orbits becomes unstable, when tidal friction will again become of overwhelming importance.

86. One of the most important and interesting parts of the solar system problem is the question of the origin of the energy which is radiated so copiously by the sun. The first idea which suggests itself to us, is that the sun's 'heat' is due to chemical action; but this will be shown to be inadmissible by considering the most powerful chemical action known, viz., the union of oxygen and hydrogen to form water.

If we take 8 grammes of oxygen and one gramme of hydrogen at $0°$ C. and a pressure of one atmo, and by first imparting and afterwards abstracting heat, cause them to form 9 grammes of water at $0°$ C. under a uniform normal pressure which remains constantly equal to one atmo during the process, the heat evolved will be about 34,000 calories, or $\dfrac{34,000}{9}$ calories per gramme of water formed.

Now if we treat the sun as a perfect sphere of r centimetres radius and uniform density ρ, and suppose that the total amount of heat which it can radiate is $\dfrac{34,000}{9}$ calories per gramme, or $\dfrac{4}{3}\pi\rho r^3 \times \dfrac{34,000}{9}$ calories in all; the total amount of heat radiated from each square centimetre of the surface, will be

$$\frac{1}{3}\rho r \times \frac{34,000}{9} \text{ calories}.$$

Taking r to be 71×10^9, and ρ, $\frac{5}{4}$, the last expression becomes

$$\frac{5 \times 71 \times 10^9}{4 \times 3} \times \frac{34{,}000}{9} \text{ calories,}$$

or $\qquad 1120 \times 10^{11}$ calories.

Hence since the radiation from each square centimetre of the sun's surface is 1215 calories per second, chemical action can only keep it up at its present rate for

$$\frac{1120 \times 10^{11}}{1215} \text{ seconds,}$$

or about 2900 years.

Chemical action being thus quite insufficient to account for the sun's 'heat,' we are driven to adopt the nebular hypothesis, as developed by Mayer, Helmholtz, and Thomson. According to this theory, the sun was originally in the form of a very attenuated gas, in which state it would possess an enormous amount of gravitational potential energy; and during the condensation of the materials to their present bulk, the original potential energy has been gradually drawn upon to supply the heat which has been so freely radiated away for countless ages. This will be made clear by the following calculations.

If two small bodies of masses m, m', change their distance apart from R to r, the work done upon them by their mutual gravitational attraction, or the decrease in their mutual gravitational potential energy, will be

$$\int_R^r \lambda \frac{mm'}{r^2}(-dr) = \lambda mm'\left(\frac{1}{r} - \frac{1}{R}\right).$$

If the masses be originally at a great distance from one

228   ELEMENTARY THERMODYNAMICS.

another, we may put R infinite, and the integral becomes simply

$$\lambda \frac{mm'}{r}.$$

Hence the gravitational potential energy lost by the materials of the sun during condensation may be written

$$\lambda \Sigma \frac{mm'}{r},$$

where Σ refers to the present state.

Denoting the distance between the two small masses m_p, m_q, by r_{pq}, we have

$$\lambda \Sigma \frac{mm'}{r} = \lambda \left\{ \frac{m_1 m_2}{r_{12}} + \frac{m_1 m_3}{r_{13}} + \frac{m_1 m_4}{r_{14}} + \ldots \right.$$
$$+ \frac{m_2 m_3}{r_{23}} + \frac{m_2 m_4}{r_{24}} + \ldots$$
$$+ \frac{m_3 m_4}{r_{34}} + \ldots$$
$$\left. + \ldots \right\}$$

$$= \tfrac{1}{2} \lambda \left\{ \frac{m_1 m_2}{r_{12}} + \frac{m_1 m_3}{r_{13}} + \frac{m_1 m_4}{r_{14}} + \ldots \right.$$
$$+ \frac{m_2 m_1}{r_{21}} + \frac{m_2 m_3}{r_{23}} + \frac{m_2 m_4}{r_{24}} + \ldots$$
$$+ \frac{m_3 m_1}{r_{31}} + \frac{m_3 m_2}{r_{32}} + \frac{m_3 m_4}{r_{34}} + \ldots$$
$$\left. + \ldots \right\}$$

Thus if we put

$$V_1 = \frac{m_2}{r_{12}} + \frac{m_3}{r_{13}} + \frac{m_4}{r_{14}} + \ldots,$$

$$V_2 = \frac{m_1}{r_{21}} + \frac{m_3}{r_{23}} + \frac{m_4}{r_{24}} + \ldots$$

we obtain

$$\lambda \Sigma \frac{mm'}{r} = \tfrac{1}{2} \lambda (m_1 V_1 + m_2 V_2 + m_3 V_3 + \ldots\ldots)$$
$$= \tfrac{1}{2} \lambda \Sigma m V.$$

We may therefore determine the value of $\lambda \Sigma \dfrac{mm'}{r}$ from the properties of the simple function V, which is such that its value at any point P, called the potential of the system at P, is $\Sigma \dfrac{m}{r}$, where r is the distance of the small mass m from P, and the summation extends to the whole material system with the exception of any small masses which may be situated indefinitely near the point P.

87. Take a thin homogeneous spherical shell of density ρ, radius a, and thickness τ, and let us find its potential at an internal point P.

Draw any line TPT' through P meeting the shell in T, T', and join T and T' to the centre C. With P as

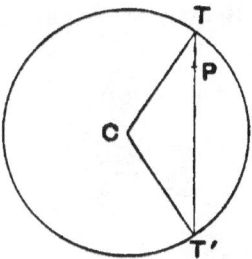

vertex and TPT' as axis, describe a small cone of vertical solid angle $d\omega$. Then if $PT = r$, the potential at P of the element cut by the cone from the shell at T, will be

$$\frac{r^2 \tau \rho d\omega}{\cos C\bar{T}T'} \cdot \frac{1}{r} = \frac{r \tau \rho d\omega}{\cos C\bar{T}T'}.$$

Similarly, if $PT' = r'$, the potential at P of the element cut from the shell at T', is

$$\frac{r'\tau\rho d\omega}{\cos CT'T}.$$

Since the angles CTT', $CT'T$, are equal, the sum of these two elementary terms is

$$\frac{(r + r')\,\tau\rho d\omega}{\cos CTT'},$$

or $\qquad\qquad 2a\tau\rho d\omega.$

Hence the potential at P of the whole shell is $4\pi a\tau\rho$, which is independent of the position of P and the same as the potential at the centre C.

To find the potential at any external point Q, take an internal point P on CQ such that

$$CQ \cdot CP = a^2.$$

Then since $\qquad\dfrac{CP}{a} = \dfrac{a}{CQ},$

if any line TPT'' be drawn through P meeting the shell

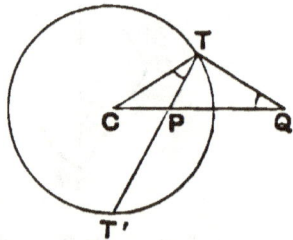

in T, T', the triangles CPT, CTQ, will be similar, and therefore

$$\frac{CQ}{TQ} = \frac{CT}{TP} \equiv \frac{a}{r}, \text{ where } PT = r.$$

Hence since the potential at Q of the element at T is

$$\frac{r^2 \tau \rho d\omega}{\cos CTT'} \cdot \frac{1}{TQ} \equiv \frac{r\tau \rho d\omega}{\cos CTT'} \cdot \frac{a}{CQ},$$

the potential of the whole shell at Q is equal to its potential at P multiplied by $\frac{a}{CQ}$, and is therefore $\frac{4\pi a^2 \tau \rho}{CQ}$, which is the same as if the whole shell had been concentrated into a particle at the centre C.

Now take a solid homogeneous shell of radius a and density ρ, and let us find its potential at an internal point P, which is at a distance x from the centre C. For this purpose, consider an elementary spherical shell of radius r and thickness dr, with its centre at C. If r be greater than x, the point P will be inside the shell and the potential of the shell at P will be $\frac{4\pi \rho r^2 dr}{r}$, or $4\pi \rho r dr$. If r be less than x, the point P will be outside the shell and the potential of the shell at P will be $\frac{4\pi \rho r^2 dr}{x}$.
Hence the potential at P of the whole sphere is

$$\int_x^a 4\pi\rho r dr + \int_0^x \frac{4\pi\rho r^2 dr}{x} \equiv 2\pi\rho a^2 - \frac{2}{3}\pi\rho x^2.$$

Since when we put a and x both zero, the last expression vanishes, we see that the potential at any point P of a small mass surrounding that point is zero. We may therefore remove the restriction that in finding the potential of a system at any point P, we are to take no account of the mass situated at that point.

88. In applying these properties of the potential, we shall treat the sun as a perfect sphere of radius a and uniform density ρ. This, of course, cannot be accurate, but it will simplify the calculations and will show the value of the nebular hypothesis.

With this assumption

$$\tfrac{1}{2}\lambda \Sigma m V = \tfrac{1}{2}\lambda \int_0^a \left(2\pi\rho a^2 - \tfrac{2}{3}\pi\rho x^2\right) 4\pi\rho x^2 dx$$

$$= \lambda \pi^2 \rho^2 \left(\tfrac{4}{3} - \tfrac{4}{15}\right) a^5$$

$$= \tfrac{16}{15} \lambda \pi^2 \rho^2 a^5.$$

If M be the whole mass, $M = \tfrac{4}{3}\pi\rho a^3$, and therefore

$$\tfrac{1}{2}\lambda \Sigma m V = \tfrac{3}{5} \lambda \tfrac{M^2}{a}.$$

Hence, taking the first form, and putting $\lambda = \tfrac{6 \cdot 5}{10^8}$, $\pi^2 = 10$, $\rho = \tfrac{5}{4}$, $a = 7 \times 10^{10}$, we find

$$\tfrac{1}{2}\lambda \Sigma m V = \tfrac{16}{15} \times \tfrac{6 \cdot 5}{10^8} \times \left(\tfrac{5}{4}\right)^2 \times 7^5 \times 10^{50}$$

$$= \frac{6 \cdot 5 \times 5 \times 7^5 \times 10^{42}}{3}$$

$$= 18 \times 10^{47}.$$

If a contract from its present value to $a\left(1 - \tfrac{1}{10,000}\right)$, we have, by the second formula for $\tfrac{1}{2}\lambda \Sigma m V$,

$$d(\tfrac{1}{2}\lambda \Sigma m V) = -\tfrac{1}{10,000} \tfrac{1}{2}\lambda \Sigma m V$$

$$= -18 \times 10^{43}.$$

Since the radiation emitted by the sun in a year is 10^{41} ergs, it follows that in the contraction of the sun to its present size, the work done by the mutual gravitation of its parts would, on our hypothesis, supply the energy radiated at the present rate in 18 million years, and a further contraction of one 10,000th part of the diameter would keep up the present rate for 1800 years.

89. The radiation forces at the surface of the sun can easily be estimated. For if we take the temperature of the sun's surface to be 10,000° C., the average non-mechanical kinetic energy of each surface particle may be taken to be $\frac{10,273}{273}$, or $37\frac{1}{2}$, times as great as it would be at 0° C. We may therefore assume the average non-mechanical velocity of each surface particle at $6 \times 50,000$ centimetres per second. Hence, since the radiation from each square centimetre of the sun's surface is 525×10^3 ergs per second, the total radiation force per square centimetre will be 175,000 dynes.

The resultant of the radiation forces acting on any finite area, found according to the usual rules of statics, will generally be quite insignificant; but since Carnot's principle does not require it to be strictly zero except in the invariable state, it is necessary to suppose that the radiation forces can have a moment about an axis through the centre of mass. As a basis for calculation, let us suppose that at any instant, all those particles which are sending out radiation, are moving in such a way that the n-th part of the radiation forces concur to oppose the angular rotation of the sun. Then the moment about the axis of rotation of the radiation forces will be

$$\frac{2}{n}\int_0^{\frac{\pi}{2}} 175{,}000 \times 2\pi a^3 \sin^2\theta\, d\theta,$$

or
$$\frac{175{,}000 \times \pi^2 a^3}{n}.$$

Now if ω be the angular velocity of the body about its axis, the angular momentum about the axis may be

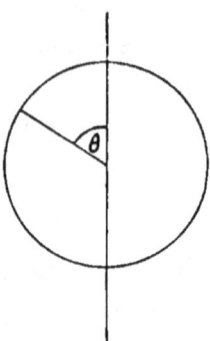

written $C\omega$, or $Mk^2\omega$, where k is the radius of gyration.

Hence
$$Mk^2 \frac{d\omega}{dt} = - \frac{175{,}000 \pi^2 a^3}{n},$$

or, putting $\frac{4}{3}\pi\rho a^3$ for M,

$$\rho k^2 \frac{d\omega}{dt} = - \frac{525}{4n} \times 10^3 \times \pi.$$

Consequently, if we assume $k = \frac{a}{3}$ and put $\frac{5}{4}$ for ρ, we find

$$\frac{d\omega}{dt} = -945 \times 10^3 \times \frac{\pi}{na^2}.$$

If this state continue for a finite time,

$$\omega = \omega_0 - 945 \times 10^3 \times \frac{\pi}{na^2} t,$$

where ω_0 is the value of ω when $t = 0$.

Thus ω will become zero in $\dfrac{na^2\omega_0}{945 \times 10^3 \times \pi}$ seconds.

Substituting for a and ω_0, we find

$$\frac{na^2\omega_0}{945 \times 10^3 \times \pi}$$

$$= n \times 7^2 \times 10^{20} \times \frac{2\pi}{25 \times 24 \times (60)^2} \times \frac{1}{945 \times 10^3 \times \pi}$$

$$= \frac{10^{13} \times n}{58 \times 36}$$

$$= \frac{10^{11}}{21} \times n.$$

Reducing to years,

$$\frac{10^{11}}{21} \times n \text{ seconds} = \frac{10^{11} \times n}{21 \times (60)^2 \times 24 \times 365} \text{ years}$$

$$= \frac{10^{11} \times n}{756 \times 10^3 \times 876} \text{ years}$$

$$= \frac{10^8 \times n}{662{,}256} \text{ years}$$

$$= 151 \times n \text{ years}.$$

From this it appears that if the sun's angular momentum has been acquired by means of the radiation forces, it must have taken a great length of time, since we cannot suppose more than a small part of the radiation forces ever to have concurred to give a mechanical resultant.

It will, of course, be seen that we cannot accept Newton's third law of motion, as usually stated, that to every force acting on a material particle there is an exactly equal and opposite force acting simultaneously on some other material particle, near or distant. Throughout this work we have assumed that action and reaction are equal and opposite in the case of two particles in contact, whether these particles are both material, or both ethereal, or one of them material and the other ethereal.

90. The sun's mechanical kinetic energy of rotation is $\frac{1}{2}C\omega^2$, or $\frac{1}{2}\frac{(C\omega)^2}{C}$. Thus if we suppose that when the sun was in the form of a very attenuated gas and C consequently very great, the angular momentum of rotation was the same as now, the mechanical kinetic of rotation would then be practically zero. Hence if we take no notice of the translation of the centre of mass and suppose the absolute temperature to have been originally zero, we may give the following brief description of the history of the sun within the period over which our present scientific knowledge extends.

Originally, the energy of the sun was chiefly gravitational potential energy, or ethereal kinetic energy bound to, or entangled among, its material particles. Then by the condensation of the mass, much of this ethereal kinetic energy has been transferred from the ether to the material particles of the sun, and by the latter radiated into space, that is, set free or lost.

If the fixed stars have no disturbing effect on the sun, an invariable state will ultimately be attained in which there will be no non-mechanical kinetic energy of matter, but in which there will be some gravitational potential energy, and also, unless Encke's hypothesis be true, mechanical kinetic energy of matter.

CHAPTER IV.

APPLICATIONS OF CARNOT'S PRINCIPLE.

91. If a material system be protected from external electric influences, the principle of the Conservation of Energy may be expressed in the form

$$dU = dQ + dW,$$

and if at every instant the temperature be uniform throughout the system and the operation reversible, Carnot's principle gives

$$dQ = \theta d\phi.$$

These two important fundamental equations will be used to investigate some of the properties of bodies of uniform temperature in which there are no electric actions and no mechanical motions. In these bodies we shall suppose that the only external forces are gravity and surface pressures; and, except when it is expressly stated otherwise, only those changes of state will be considered in which no mechanical work is done on the body except by a uniform normal surface pressure. Denoting this pressure by p and the volume of the body by v, we have

$$dW = -pdv,$$

and if the operation be reversible,
$$dU = \theta d\phi - p dv.$$
When, as frequently happens, any two of the five quantities (θ, p, v, U, ϕ) can be taken as independent variables to define the state of the body, we obtain also
$$d(U + pv) = \theta d\phi + v dp,$$
and
$$\left. \begin{array}{l} d(U - \theta\phi) = -\phi d\theta - p dv \\ d(U - \theta\phi + pv) = -\phi d\theta + v dp \end{array} \right\}.$$
Hence by expressing that these quantities are complete differentials, we get

$$\left. \begin{array}{l} \dfrac{d_\phi \theta}{dv} = -\dfrac{d_v p}{d\phi}, \\[6pt] \dfrac{d_\phi \theta}{dp} = \dfrac{d_p v}{d\phi}, \\[6pt] \dfrac{d_\theta \phi}{dv} = \dfrac{d_v p}{d\theta}, \\[6pt] \dfrac{d_\theta \phi}{dp} = -\dfrac{d_p v}{d\theta}. \end{array} \right\} \quad \ldots\ldots\ldots\ldots\ldots(52).$$

These results are known as 'the four thermodynamic relations.'

92. The problems we propose to discuss in the present chapter may be divided into three classes. In the first, the body is supposed to be homogeneous throughout, like a gas or a piece of iron or india-rubber. In the second, it is supposed to consist of two or more homogeneous parts, which are alike in substance but differ from one another in physical state, as in the case of water and steam, or ice and water. In the third, we shall chiefly consider bodies consisting of two or more homogeneous parts which are not alike in substance. As an illustrative example, we may take the case of an aqueous solution of a salt with a quantity of salt undissolved.

PART I.

ON HOMOGENEOUS BODIES.

93. We shall first suppose that when the substance is in a state of equilibrium, we may take any two of the three quantities (θ, p, v) as independent variables, and treat the third as a dependent variable. Also, for simplicity, we shall take the mass of the body to be one gramme.

Then if (θ, v) be chosen as independent variables,

$$dU = \frac{d_v U}{d\theta} d\theta + \frac{d_\theta U}{dv} dv.$$

But if $C_v d\theta$ be the heat required to raise the temperature of the body from θ to $\theta + d\theta$, while the volume is kept constant, we have

$$C_v d\theta = \frac{d_v U}{d\theta} d\theta,$$

and therefore

$$dU = C_v d\theta + \frac{d_\theta U}{dv} dv \quad \ldots\ldots\ldots\ldots(53).$$

We shall define C_v to be the 'specific heat of the substance at constant volume.' This definition is not quite the same as that given in Art. 33; but, as a matter of fact, the specific heat varies so little with the temperature that the two definitions are identical for experimental purposes.

Again, we have
$$dU = dQ - p\,dv.$$
Hence
$$dQ = C_v d\theta + \left(\frac{d_\theta U}{dv} + p\right) dv,$$
which may be writtten
$$dQ = C_v d\theta + \frac{d_\theta Q}{dv} dv,$$
................(54),

so that
$$\frac{d_\theta Q}{dv} = \frac{d_\theta U}{dv} + p.$$

This expression for dQ only holds for a reversible modification, because we have supposed the body to be always in a state of equilibrium.

To find $\dfrac{d_\theta Q}{dv}$, let the substance be made to undergo the following reversible cycle of operations.

(1) Let the volume be slowly increased from v to $v + dv$, whilst the temperature is kept constantly equal to θ. The heat absorbed will be $\dfrac{d_\theta Q}{dv} dv$.

(2) Let the state of the substance then be altered without loss or gain of heat so that the temperature falls to $\theta - \tau$, where τ is indefinitely small.

(3) Let the volume now be slowly diminished at the constant temperature $\theta - \tau$ by such an amount that in the fourth operation in which there is neither loss nor gain of heat, it may be possible to restore the substance to its original state.

On the indicator diagram, the cycle will be represented by a small parallelogram $ABCD$, whose area will be equal

APPLICATIONS OF CARNOT'S PRINCIPLE. 241

to that of $ABba$, where Aa, Bb are drawn parallel to Op. Hence, since the abscissae of A and B are v and $v+dv$,

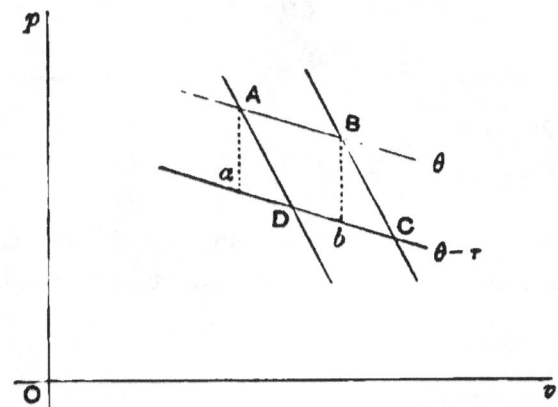

respectively, the work done by the substance will be $Aa \cdot dv$, or $\tau \dfrac{d_v p}{d\theta} dv$.

But we have already seen, in Art. 51, that the efficiency of the cycle is $\dfrac{\tau}{\theta}$. Hence

$$\frac{\tau \dfrac{d_v p}{d\theta} dv}{\dfrac{d_\theta Q}{dv} dv} = \frac{\tau}{\theta},$$

or

$$\frac{d_\theta Q}{dv} = \theta \frac{d_v p}{d\theta} \quad \ldots\ldots\ldots\ldots\ldots\ldots(55).$$

This remarkable result was practically obtained by Carnot, who found

$$\frac{d_\theta Q}{dv} = C \frac{d_v p}{d\theta},$$

C being an unknown function, now called Carnot's function, which was independent of the nature of the working substance.

P.

Substituting in equations (53) and (54), we get

$$dU = C_v d\theta + \left(\theta \frac{d_v p}{d\theta} - p\right) dv,$$

$$dQ = C_v d\theta + \theta \frac{d_v p}{d\theta} dv, \qquad \bigg\} \dots\dots\dots(56).$$

and therefore

$$d\phi = C_v \frac{d\theta}{\theta} + \frac{d_v p}{d\theta} dv.$$

Expressing the condition either that dU or that $d\phi$ is a complete differential, we obtain the curious result

$$\frac{d_\theta C_v}{dv} = \theta \frac{d^2_v p}{d\theta^2} \dots\dots\dots\dots(57),$$

whatever the substance may be.

Again, if $C_p d\theta$ be the heat required to raise the temperature of the body from θ to $\theta + d\theta$, C_p may be called the specific heat of the body at constant pressure, and the fundamental equation $dU = dQ - pdv$ becomes, when p is constant,

$$dU = C_p d\theta - pdv,$$

or
$$C_p d\theta = dU + pdv$$
$$= d(U + pv),$$

or
$$C_p = \frac{d_p(U + pv)}{d\theta}.$$

Hence if we take θ and p as independent variables, we have in any reversible operation,

$$dQ = dU + pdv$$
$$= d(U + pv) - vdp$$
$$= \frac{d_p(U+pv)}{d\theta} d\theta + \left\{\frac{d_\theta(U+pv)}{dp} - v\right\} dp$$
$$= C_p d\theta + \frac{d_\theta Q}{dp} dp, \text{ say.}$$

APPLICATIONS OF CARNOT'S PRINCIPLE. 243

To find $\dfrac{d_\theta Q}{dp}$, we take the same reversible cycle as before, and draw Aa', Bb' parallel to Ov to meet CD.

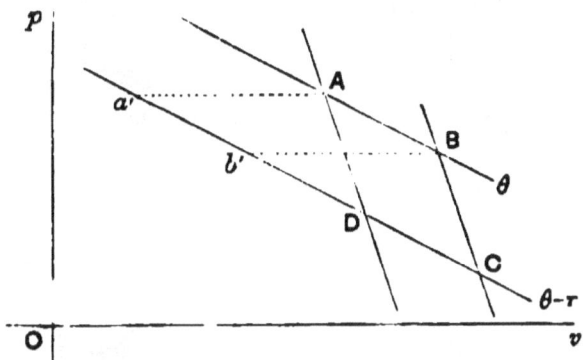

Hence the area of $ABCD$ may be written
$$-Aa' \cdot dp,$$
for the ordinates of A and B are p and $p + dp$ respectively. Substituting $\tau \dfrac{d_p v}{d\theta}$ for Aa', this becomes
$$-\tau \frac{d_p v}{d\theta} dp.$$

The heat absorbed during the operation represented by AB being $\dfrac{d_\theta Q}{dp} dp$, we have therefore

$$\frac{-\tau \dfrac{d_p v}{d\theta} dp}{\dfrac{d_\theta Q}{dp} dp} = \frac{\tau}{\theta},$$

or
$$\frac{d_\theta Q}{dp} = -\theta \frac{d_p v}{d\theta} \quad\ldots\ldots\ldots\ldots\ldots(58).$$

Hence
$$dQ = C_p d\theta - \theta \frac{d_p v}{d\theta} dp,$$
$$d\phi = C_p \frac{d\theta}{\theta} - \frac{d_p v}{d\theta} dp, \quad \quad \ldots\ldots(59);$$
$$dU = dQ - p\,dv$$
$$= C_p d\theta - \left(\theta \frac{d_p v}{d\theta} - v\right) dp - d(pv),$$

and since dU and $d\phi$ are complete differentials,
$$\frac{d_\theta C_p}{dp} = -\theta \frac{d^2_p v}{d\theta^2} \quad \ldots\ldots\ldots(60).$$

94. If, in equation (56), we suppose that θ and v vary in such a manner that p remains constant, we shall have $dQ = C_p d\theta$ and therefore
$$C_p = C_v + \theta \frac{d_v p}{d\theta} \frac{d_p v}{d\theta}.$$

But since
$$dp = \frac{d_v p}{d\theta} d\theta + \frac{d_\theta p}{dv} dv,$$

we have
$$0 = \frac{d_v p}{d\theta} + \frac{d_\theta p}{dv} \frac{d_p v}{d\theta}.$$

Hence
$$C_p - C_v = -\theta \frac{\left(\frac{d_v p}{d\theta}\right)^2}{\left(\frac{d_\theta p}{dv}\right)} \quad \ldots\ldots(61).$$

If θ and p be taken as independent variables, we shall obtain
$$C_p - C_v = -\theta \frac{\left(\frac{d_p v}{d\theta}\right)^2}{\left(\frac{d_\theta v}{dp}\right)} \quad \ldots\ldots(62).$$

APPLICATIONS OF CARNOT'S PRINCIPLE.   245

For the ideal perfect gas, $pv = R\theta$, so that

$$\frac{d_v p}{d\theta} = \frac{R}{v}, \quad \frac{d_p v}{d\theta} = \frac{R}{p}, \quad \frac{d_\theta p}{dv} = -\frac{R\theta}{v^2},$$

and therefore, by equations (57), (60), and (61), C_p and C_v are functions of θ only such that

$$C_p - C_v = R.$$

Also, by equations (56),

$$\left. \begin{array}{l} dU = C_v d\theta \\ dQ = C_v d\theta + p\, dv \end{array} \right\}.$$

Hence also, since dU is a complete differential, we see that both C_v and U can only vary with θ, just as in Chap. II.

95. If the substance at pressure p, volume v, and temperature θ, be compressed by an increase of pressure dp to volume $v + dv$, the Isothermal Compressibility, K_θ, is defined to be

$$\left. \begin{array}{l} K_\theta = \dfrac{-\dfrac{d_\theta v}{v}}{d_\theta p} = -\dfrac{1}{v}\dfrac{d_\theta v}{dp}, \\[2ex] K_\phi = \dfrac{-\dfrac{d_\phi v}{v}}{d_\phi p} = -\dfrac{1}{v}\dfrac{d_\phi v}{dp}. \end{array} \right\} \ldots\ldots\ldots\ldots(63).$$

and the Adiabatic Compressibility, K_ϕ,

The reciprocal of the compressibility may be called the Elasticity of volume, and be denoted by E: thus

$$\left. \begin{array}{l} E_\theta = -v\dfrac{d_\theta p}{dv} \\[2ex] E_\phi = -v\dfrac{d_\phi p}{dv} \end{array} \right\} \ldots\ldots\ldots\ldots(64),$$

and therefore
$$\frac{E_\phi}{E_\theta} = \frac{\dfrac{d_\phi p}{dv}}{\dfrac{d_\theta p}{dv}}.$$

But we have
$$dp = \frac{d_\phi p}{d\theta} d\theta + \frac{d_\theta p}{d\phi} d\phi,$$

so that, by the thermodynamic relations
$$\frac{d_\phi p}{dv} = \frac{d_\phi p}{d\theta}\frac{d_\phi \theta}{dv},$$

$$= -\frac{\dfrac{d_p \phi}{dv}}{\dfrac{d_v \phi}{dp}},$$

and
$$\frac{d_\theta p}{dv} = \frac{d_\theta p}{d\phi}\frac{d_\theta \phi}{dv}$$

$$= -\frac{\dfrac{d_v p}{d\theta}}{\dfrac{d_p v}{d\theta}}.$$

Hence
$$\frac{E_\phi}{E_\theta} = \frac{\dfrac{d_p \phi}{dv} \cdot \dfrac{d_p v}{d\theta}}{\dfrac{d_v \phi}{dp} \cdot \dfrac{d_v p}{d\theta}}$$

$$= \frac{\dfrac{d_p \phi}{d\theta}}{\dfrac{d_v \phi}{d\theta}},$$

since from the equation
$$d\phi = \frac{d_v \phi}{dp} dp + \frac{d_p \phi}{dv} dv,$$

APPLICATIONS OF CARNOT'S PRINCIPLE.

we get
$$\frac{d_p\phi}{d\theta} = \frac{d_p\phi}{dv} \cdot \frac{d_pv}{d\theta},$$

and
$$\frac{d_v\phi}{d\theta} = \frac{d_v\phi}{dp} \cdot \frac{d_vp}{d\theta}.$$

Now from the equation $dQ = \theta d\phi$, we find

$$\left. \begin{array}{l} C_p = \theta \dfrac{d_p\phi}{d\theta} \\[1ex] C_v = \theta \dfrac{d_v\phi}{d\theta} \end{array} \right\}.$$

Thus we have the important relation
$$\frac{E_\phi}{E_\theta} = \frac{C_p}{C_v} \equiv k \ldots\ldots\ldots\ldots\ldots(65).$$

Again, we define e, the 'coefficient of cubical dilatation by heat at constant pressure', to be such that
$$\frac{d_pv}{v} = ed_p\theta,$$

or
$$e = \frac{1}{v}\frac{d_pv}{d\theta} \ldots\ldots\ldots\ldots\ldots\ldots(66).$$

But from the relation
$$dp = \frac{d_vp}{d\theta} d\theta + \frac{d_\theta p}{dv} dv$$

we obtain
$$0 = \frac{d_vp}{d\theta} + \frac{d_\theta p}{dv}\frac{d_pv}{d\theta}.$$

Hence
$$e = -\frac{1}{v}\frac{\dfrac{d_vp}{d\theta}}{\dfrac{d_\theta p}{dv}} = \frac{1}{E_\theta}\frac{d_vp}{d\theta} \ldots\ldots\ldots\ldots(67).$$

In terms of these definitions, we obtain
$$C_p - C_v = \theta E_\theta e^2 v \ldots\ldots\ldots\ldots\ldots(68).$$

For the ideal perfect gas, $pv = R\theta$, and therefore

$$e = \frac{1}{\theta}$$
$$E_\theta = p$$
$$E_\phi = \frac{C_p}{C_v} p$$

For liquids and solids at constant pressure, we may generally write

$$v = v'(1 + \alpha\theta'),$$

for moderately small changes of temperature, where

$$\theta' = \theta - 273,$$

v' is the value of v at $0°$ C., and α a small number independent of θ.

Consequently, $\quad v = v'(1 - 273\alpha)(1 + \alpha\theta),$

and $\quad \log v = \log[v'(1 - 273\alpha)] + \alpha\theta$, nearly.

Hence $\quad e = \dfrac{1}{v}\dfrac{d_p v}{d\theta} = \alpha$, a function of p only.

Solid bodies, in expanding by heat at constant pressure, preserve their forms. Thus if l be the length of the edge of a cube at temperature θ, we may write

$$l = l'(1 + \beta\theta'),$$

where l' is the value of l at $0°$ C., and

$$1 + \alpha\theta' = (1 + \beta\theta')^3,$$

or $\quad\quad\quad\quad\quad \alpha = 3\beta.$

The small number β is called the 'coefficient of linear expansion.'

96. The value of equation (68) may be illustrated by the case of water; for C_v cannot be determined for water

APPLICATIONS OF CARNOT'S PRINCIPLE.

by direct experiment. Thus if J be the number of ergs in a calorie, or 'Joule's equivalent,' we have at a pressure of one atmo:

$$\left.\begin{array}{l}\text{at } 0°\text{ C.},\ldots C_p = J,\ldots e = -\cdot000057,\ldots v = 1.\\ \text{at } 25°\text{ C.},\ldots C_p = J \times 1\cdot 0016,\ldots e = \cdot 00022,\ldots v = 1\cdot 003.\\ \text{at } 50°\text{ C.},\ldots C_p = J \times 1\cdot 0042,\ldots e = \cdot 00049,\ldots v = 1\cdot 012.\end{array}\right\}$$

Also when the pressure is one atmo, the diminution of volume due to an increase of pressure of one atmo is found to bear to the original volume the ratio:

$$\left.\begin{array}{l}\text{at } 0°\text{ C.}\ldots\ldots\cdot 00005.\\ \text{at } 25°\text{ C.}\ldots\ldots\cdot 000046.\\ \text{at } 50°\text{ C.}\ldots\ldots\cdot 000044.\end{array}\right\}$$

Hence if n be the number of dynes per square centimetre in a pressure of one atmo, we obtain, for a pressure of one atmo:

$$\left.\begin{array}{l}\text{at } 0°\text{ C.}\ldots\ldots E_\theta = \dfrac{n}{\cdot 00005} = 2n \times 10^4.\\[4pt] \text{at } 25°\text{ C.}\ldots\ldots E_\theta = \dfrac{n}{\cdot 000046} = 2\cdot 174n \times 10^4.\\[4pt] \text{at } 50°\text{ C.}\ldots\ldots E_\theta = \dfrac{n}{\cdot 000044} = 2\cdot 273n \times 10^4.\end{array}\right\}$$

Thus at a pressure of one atmo, we have for water:

$$\left.\begin{array}{l}\text{at } 0°\text{ C.}\ldots\ldots C_p - C_v = J \times \cdot 00044,\\ \text{at } 25°\text{ C.}\ldots\ldots C_p - C_v = J \times \cdot 0075,\\ \text{at } 50°\text{ C.}\ldots\ldots C_p - C_v = J \times \cdot 0425,\end{array}\right\}$$

and therefore:

$$\left.\begin{array}{l}\text{at } 0°\text{ C.}\ldots\ldots C_v = J \times \cdot 99956.\\ \text{at } 25°\text{ C.}\ldots\ldots C_v = J \times \cdot 9941.\\ \text{at } 50°\text{ C.}\ldots\ldots C_v = J \times \cdot 9617.\end{array}\right\}$$

97. When the substance undergoes a reversible operation in which heat is neither gained nor lost, the equation
$$dQ = C_v d\theta + \theta \frac{d_v p}{d\theta} dv$$
gives
$$d\theta = -\frac{\theta e E_\theta}{C_v} dv \dots\dots\dots\dots\dots(69).$$

If we take the equation
$$dQ = C_p d\theta - \theta \frac{d_p v}{d\theta} dp,$$
we get
$$d\theta = \frac{\theta e v}{C_p} dp \dots\dots\dots\dots\dots(70).$$

If then, when the pressure is kept constant, the substance contracts as its temperature rises, like water between the freezing point and its point of maximum density, e will be negative, and in an isentropic operation, $d\theta$ and dp will have opposite signs, so that an increase of pressure will cause a fall of temperature.

98. We will now explain the experiment by which Joule and Thomson determined more exactly the absolute temperature of the freezing point and the true law of what are called perfect gases.

A stream of gas is kept constantly flowing by means of a pump through a long pipe in one short length of which there is firmly fixed a porous plug of cotton-wool or waste silk, by which the motion of the gas is so impeded that its velocity remains small even when there is a considerable difference between the pressures before and behind the plug. If the pump be at a sufficient distance and worked as steadily as possible, the pulsations which it causes will be imperceptible, and after a little time from the commencement of the experiment, the state of the motion will become steady except in the immediate neighbourhood of the plug.

Suppose then that P and Q are two sections, one before and the other behind the plug, but at such a distance from it that the irregular motions and pressures due

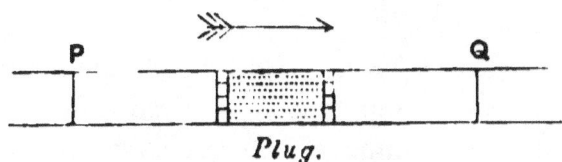

Plug.

to the passage through it are not discernible, so that there is only a uniform current of gas to be considered. And let the portion PQ of the pipe be surrounded by some non-conducting material, so that heat can neither enter nor escape through it.

Let p and θ be the constant pressure and temperature and v the constant volume per gramme, of the gas which passes through the section P, and let (p', v', θ') be the values of the same quantities at the section Q. Then since we may imagine that the entering stream is forced towards the plug by an ideal piston at P, and that the emergent stream forces out another ideal piston at Q, it is clear that when one gramme of gas passes through the portion PQ of the pipe, the external work done upon it will be

$$pv - p'v'.$$

But, since the energy of the mechanical motion is practically zero, the change of energy is given by the equation

$$U' - U = \int_P^Q \left\{ C_p d\theta - \left(\theta \frac{d_p v}{d\theta} - v \right) dp \right\} - (p'v' - pv).$$

Hence the heat absorbed by the gas is

$$\int_P^Q \left\{ C_p d\theta - \left(\theta \frac{d_p v}{d\theta} - v\right) dp \right\},$$

and as in the experiment this is zero, we have

$$\int_P^Q \left(\theta \frac{d_p v}{d\theta} - v\right) dp = \int_P^Q C_p d\theta.$$

Now the expression on the right hand side is merely the quantity of heat in ergs that is required to raise the temperature of one gramme of the gas at constant pressure from θ to θ', and its determination does not require us to possess any previous knowledge of absolute temperature. In order to find it, the difference of temperature between the sections P and Q was measured by Joule and Thomson by a mercury thermometer, but the very same thermometer had been previously used by Joule in determining the value of J, that is, the number of ergs in a calorie. The value of C_p, or the specific heat in calories multiplied by J, was deduced from Regnault's experiments. The product was the integral $\int_P^Q C_p d\theta$, since $\theta' - \theta$ was small.

The experiments showed that for a given value of θ, the integral was simply proportional to $p - p'$, not merely for an infinitesimal difference of pressure, but for differences of 5 or 6 atmos.

In the case of hydrogen, the gas was slightly heated by passing through the plug and the heating effect was observed at temperatures from 4° C. or 5° C. to about 90° C. The investigation, however, was not carried out in sufficient detail to give any law of variation of this small effect with temperature, and we shall therefore

APPLICATIONS OF CARNOT'S PRINCIPLE.

take the mean of the results, which may be expressed thus:

$$\int_P^Q c_p d\theta = a'(p - p'),$$

where c_p is the specific heat in calories, and a' the constant number ·000 000 13116. In other words,

$$\int_P^Q C_p d\theta = a(p - p'),$$

where $a = ·000\ 000\ 13116 \times J = 5·448$.

Thus
$$\int_P^Q \left(\theta \frac{d_p v}{d\theta} - v\right) dp = a(p - p'),$$

and therefore
$$\theta \frac{d_p v}{d\theta} - v = -a \quad \ldots\ldots\ldots\ldots\ldots(71).$$

Hence
$$\frac{d_p}{d\theta}\left(\frac{v}{\theta}\right) = -\frac{a}{\theta^2},$$

or
$$v = a + \theta f(p) \quad \ldots\ldots\ldots\ldots\ldots(72),$$

where $f(p)$ is an unknown function of p.

From this equation we have

$$pv = ap + \theta p f(p).$$

Now we know that when the temperature is high enough and the pressure not too great, every perfect gas satisfies very approximately a relation of the form $pv = R\theta$. The quantity a cannot therefore be strictly constant, as we have supposed. But it appears from Joule and Thomson's experimental results that it varies little between the temperatures 4° C. and 90° C.

If the absolute temperature of the freezing point be denoted by θ_0, the absolute temperature of the boiling point will be $\theta_0 + 100$. Hence, if v_0 and v_{100} be the corresponding values of v at any the same pressure p, we shall have

$$v_0 - a = \theta_0 f'(p),$$

and
$$v_{100} - v_0 = 100 f'(p).$$

Therefore
$$\frac{v_0 - a}{v_{100} - v_0} = \frac{\theta_0}{100},$$

that is,
$$\frac{\theta_0}{100} = \frac{v_0}{v_{100} - v_0}\left(1 - \frac{a}{v_0}\right),$$

or
$$\theta_0 = \frac{100}{E}\left(1 - \frac{a}{v_0}\right) \quad \dots\dots\dots\dots(73),$$

where E^1 stands for $\dfrac{v_{100} - v_0}{v_0}$.

Now for hydrogen expanding at the constant pressure of one atmo, Regnault found $E = \cdot 36613$, and therefore $\dfrac{100}{E} = 273\cdot 13$. Also, at the pressure of one atmo, $v_0 = 11164\cdot 45$. Thus we get

$$\theta_0 = 273\cdot 13 \left(1 - \frac{5\cdot 448}{11164\cdot 45}\right)$$
$$= 273\cdot 13\, (1 - \cdot 000488)$$
$$= 273.$$

In the case of common air and carbonic acid, the thermal effect observed was a slight lowering of temperature, which was shown, in 1862, to vary at different temperatures very nearly in the inverse ratio of the square of $C + 273$, where C is the temperature at either the section P or the section Q, as shown by the mercury thermometer. Thus we have

$$\int_P^Q C_p d\theta = -b\left(\frac{273}{\theta}\right)^2 (p - p'), \text{ or } = -b\left(\frac{273}{\theta'}\right)^2 (p - p'),$$

---

[1] E must not be confused with E_θ or E_ϕ.

APPLICATIONS OF CARNOT'S PRINCIPLE.

that is,

$$\int_P^Q \left(\theta \frac{d_p v}{d\theta} - v\right) dp = -b \left(\frac{273}{\theta}\right)^2 (p - p'),$$

$$\text{or} = -b \left(\frac{273}{\theta'}\right)^2 (p - p').$$

From either of these, remembering that the variation of θ with p is small, we get

$$\theta \frac{d_p v}{d\theta} - v = b \left(\frac{273}{\theta}\right)^2,$$

and therefore

$$\frac{d_p}{d\theta}\left(\frac{v}{\theta}\right) = b \left(\frac{273}{\theta^2}\right)^2.$$

Hence

$$v = -\frac{b}{3}\left(\frac{273}{\theta}\right)^2 + \theta f(p) \quad \ldots \ldots \ldots \ldots (74),$$

and as before

$$\theta_0 f(p) = v_0 + \frac{b}{3}\left(\frac{273}{\theta_0}\right)^2,$$

$$100 f(p) = v_{100} - v_0 + \frac{b}{3}\left(\frac{273}{\theta_0}\right)^2 \left\{\left(\frac{\theta_0}{\theta_0 + 100}\right)^2 - 1\right\},$$

or

$$\theta_0 f(p) = v_0 \left\{1 + \frac{b}{3v_0}\left(\frac{273}{\theta_0}\right)^2\right\},$$

$$100 f(p) = (v_{100} - v_0) \left[1 + \frac{b}{3v_0} \frac{v_0}{v_{100} - v_0} \left(\frac{273}{\theta_0}\right)^2 \right.$$

$$\left. \times \left\{\left(\frac{\theta_0}{\theta_0 + 100}\right)^2 - 1\right\}\right],$$

so that, if $E \equiv \dfrac{v_{100} - v_0}{v_0}$,

$$\frac{E \theta_0}{100} = 1 + \frac{b}{3v_0}\left(\frac{273}{\theta_0}\right)^2 \left[1 - \frac{1}{E}\left\{\left(\frac{\theta_0}{\theta_0 + 100}\right)^2 - 1\right\}\right].$$

In the small terms on the right hand side of this equation we may put $\theta_0 = 273$. We have then the formula

$$\theta_0 = \frac{100}{E}\left\{1 + \frac{b}{3v_0}\left(1 + \frac{\cdot 464}{E}\right)\right\} \quad \ldots \ldots \ldots (75).$$

256   ELEMENTARY THERMODYNAMICS.

Now for common air, $b = 2·684$; and at the pressure of one atmo, $E = ·36706$, and $v_0 = 773·3$. The corresponding quantities for carbonic acid are $b = 12·323$, $E = ·37100$, and $v_0 = 505·7$. Thus we obtain in the two cases,

$$\theta_0 = 272·44 (1 + ·00261),$$

and

$$\theta_0 = 269·5 (1 + ·0182),$$

that is,

$$\theta_0 = 272·44 + ·71 = 273·15,$$

and

$$\theta_0 = 269·5 + 4·90 = 274·4,$$

respectively.

The results obtained from the three gases are collected in the accompanying table.

Gas.	E	Uncorrected estimate of absolute temperature of the freezing point.	Correction.	Result.
Hydrogen	·36613	273°·13	−·13°	273°·00
Air	·36706	272·44	+·71	273·15
Carbonic acid	·37100	269·5	+4·90	274·40.

99. Again, from equation (74), we have

$$pv = \theta p f(p) - \tfrac{1}{3}bp \left(\frac{273}{\theta}\right)^2.$$

But since, when the temperature is high enough, the gas satisfies the relation $pv = R\theta$, it is evident that $pf(p) = R$, a constant. Thus, finally,

$$pv = R\theta - \tfrac{1}{3}bp \left(\frac{273}{\theta}\right)^2 \quad \ldots\ldots\ldots\ldots(76).$$

This equation, according to Thomson and Joule, must be used instead of the simpler relation $pv = R\theta$.

APPLICATIONS OF CARNOT'S PRINCIPLE.

An exactly similar equation had been obtained by Rankine in 1854, in the form

$$pv = R\theta - \frac{a}{\theta v} \quad \ldots\ldots\ldots\ldots(77),$$

where a is a constant.

By substituting $\frac{R\theta}{v}$ for p in the small term of equation (76), the two will easily be seen to be identical.

From equation (76), we find

$$\frac{d_p v}{d\theta} = \frac{R}{p} + \tfrac{2}{3}\frac{b}{\theta}\left(\frac{273}{\theta}\right)^2,$$

and
$$\frac{d^2_p v}{d\theta^2} = -2b\left(\frac{273}{\theta^2}\right)^2.$$

Also since
$$pv\left\{1 + \frac{b}{3v}\left(\frac{273}{\theta}\right)^2\right\} = R\theta,$$

or
$$pv = R\theta\left\{1 - \frac{b}{3v}\left(\frac{273}{\theta}\right)^2\right\} \quad \ldots\ldots\ldots(78),$$

we obtain
$$v\frac{d_v p}{d\theta} = R + \frac{bR}{3v}\left(\frac{273}{\theta}\right)^2,$$

and
$$v\frac{d^2_v p}{d\theta^2} = -\tfrac{2}{3}\frac{bR}{v\theta}\left(\frac{273}{\theta}\right)^2.$$

Hence
$$\left.\begin{array}{l}\dfrac{d_\theta C_v}{dv} \equiv \theta\dfrac{d^2_v p}{d\theta^2} = -\tfrac{2}{3}\dfrac{bR}{v^2}\left(\dfrac{273}{\theta}\right)^2 \equiv -\tfrac{2}{3}\dfrac{bp^2}{R}\left(\dfrac{273}{\theta^2}\right)^2 \\ \dfrac{d_\theta C_p}{dp} \equiv -\theta\dfrac{d^2_p v}{d\theta^2} = \dfrac{2b}{\theta}\left(\dfrac{273}{\theta}\right)^2\end{array}\right\}\ldots(79).$$

As an illustration, let c_p be the specific heat at constant pressures in calorics, P the pressure in atmos, J the number of ergs in a calorie, and n the number of dynes per square centimetre in a pressure of one atmo. Then

$$\frac{d_\theta c_p}{dP} = \frac{n}{J}\frac{d_\theta C_p}{dp} = \frac{2bn}{J\theta}\left(\frac{273}{\theta}\right)^2.$$

Thus for common air,

$$\frac{d_\theta c_p}{dP} = \frac{\cdot 1308}{\theta} \cdot \left(\frac{273}{\theta}\right)^2,$$

and for carbonic acid,

$$\frac{d_\theta c_p}{dP} = \frac{\cdot 6}{\theta} \cdot \left(\frac{273}{\theta}\right)^2.$$

Again,
$$e = \frac{1}{v}\frac{d_p v}{d\theta} = \frac{R}{pv} + \frac{2b}{3v\theta}\left(\frac{273}{\theta}\right)^2$$
$$= \frac{1}{\theta} + \frac{b}{v\theta}\left(\frac{273}{\theta}\right)^2 \quad\ldots\ldots\ldots\ldots(80).$$

Also
$$E_\theta = -v\frac{d_\theta p}{dv} = p - \frac{b}{3v^2}\left(\frac{273}{\theta}\right)^2 R\theta$$
$$= p\left\{1 - \frac{b}{3v}\left(\frac{273}{\theta}\right)^2\right\} \quad\ldots\ldots(81),$$

and
$$C_p - C_v = \theta E_\theta e^2 v$$
$$= \frac{pv}{\theta}\left\{1 - \frac{b}{3v}\left(\frac{273}{\theta}\right)^2\right\}\left\{1 + \frac{2b}{v}\left(\frac{273}{\theta}\right)^2\right\}$$
$$= R\left\{1 + \frac{4b}{3v}\left(\frac{273}{\theta}\right)^2\right\} \quad\ldots\ldots\ldots\ldots\ldots(82).$$

100. As a further illustration of Carnot's principle, we will consider the case of a strained rod, the straining force being so great that the pressure of the atmosphere and the weight of the rod may be neglected.

Let T be the force, considered positive when tensional, l the length of the rod when acted on by the force and θ

APPLICATIONS OF CARNOT'S PRINCIPLE. 259

its uniform temperature. Then in any reversible process, taking θ and T as independent variables, we have

$$dQ = \frac{d_T Q}{d\theta} d\theta + \frac{d_\theta Q}{dT} dT$$

$$= MC_T d\theta + \frac{d_\theta Q}{dT} dT,$$

where M is the mass of the rod and C_T its specific heat under constant tension.

To find $\frac{d_\theta Q}{dT}$, let the rod be made to undergo the following complete cycle of reversible operations.

(1) Let the rod be slowly stretched at constant temperature until T increases by dT. The heat absorbed will be $\frac{d_\theta Q}{dT} dT$.

(2) Let the rod be still further stretched, but without loss or gain of heat, until the temperature becomes $\theta - \tau$, where τ is indefinitely small.

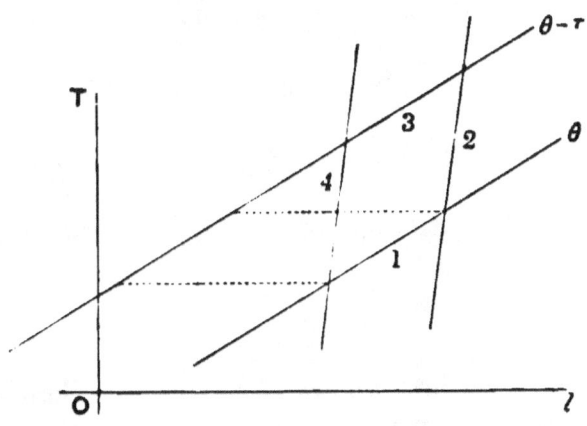

(3) Then, whilst the temperature is kept constantly equal to $\theta - \tau$, let the tensions be slowly reduced by such

17—2

an amount that an adiabatic operation will restore the rod to its original state.

If the cycle be represented by a diagram in which T and l are independent variables, it will easily be seen that the work done by the rod during the cycle, is

$$\tau \frac{d_\tau l}{d\theta} dT.$$

We have therefore

$$\frac{\tau \dfrac{d_\tau l}{d\theta} dT}{\dfrac{d_\theta Q}{dT} dT} = \frac{\tau}{\theta},$$

or
$$\frac{d_\theta Q}{dT} = \theta \frac{d_\tau l}{d\theta} \dots\dots\dots\dots\dots(83).$$

Thus
$$dQ = MC_\tau d\theta + \theta \frac{d_\tau l}{d\theta} dT \dots\dots\dots(84),$$

and therefore, since $dQ = \theta d\phi$,

$$d\phi = MC_\tau \frac{d\theta}{\theta} + \frac{d_\tau l}{d\theta} dT \dots\dots\dots(85).$$

The condition that $d\phi$ is a complete differential, is

$$\frac{d_\theta (MC_\tau)}{dT} = \theta \frac{d^2_\tau l}{d\theta^2} \dots\dots\dots\dots(86).$$

In any adiabatic operation, dQ and $d\phi$ are both zero: hence

$$\frac{d_\phi \theta}{dT} = - \frac{\theta}{MC_\tau} \frac{d_\tau l}{d\theta} \dots\dots\dots\dots(87).$$

This relation was first obtained by Sir W. Thomson and led him to make a curious prediction with respect to the behaviour of india-rubber, which was experimentally verified by Joule in 1859.

APPLICATIONS OF CARNOT'S PRINCIPLE.

So long as india-rubber is acted on by no force or only by a small force, it exhibits the same phenomena as most other substances, lengthening when heated and shortening when cooled. But when the force is great enough, it shortens when heated and lengthens when cooled. Under these circumstances, $\frac{d_r l}{d\theta}$ is negative, and therefore, since C_r is positive, we see that $\frac{d_\phi \theta}{dT}$ must be positive. India-rubber will therefore be heated by an increase of the straining force when that force is large enough, and conversely.

Again, equation (84) may be written

$$dQ = MC_r d\theta + \theta \frac{d_r l}{d\theta} \left(\frac{d_l T}{d\theta} d\theta + \frac{d_\theta T}{dl} dl \right)$$

$$= \left(MC_r + \theta \frac{d_r l}{d\theta} \frac{d_l T}{d\theta} \right) d\theta + \theta \frac{d_r l}{d\theta} \frac{d_\theta T}{dl} dl.$$

Hence, since
$$0 = \frac{d_r l}{d\theta} + \frac{d_\phi l}{dT} \frac{d_l T}{d\theta} \quad \ldots\ldots\ldots\ldots\ldots\ldots(88),$$

we have

$$dQ = \left(MC_r + \theta \frac{d_r l}{d\theta} \frac{d_l T}{d\theta} \right) d\theta - \theta \frac{d_l T}{d\theta} dl.$$

Thus if we denote by C_l the specific heat of the rod at constant length, we obtain

$$\left. \begin{array}{l} dQ = MC_l d\theta - \theta \dfrac{d_l T}{d\theta} dl \\[6pt] d\phi = MC_l \dfrac{d\theta}{\theta} - \dfrac{d_l T}{d\theta} dl \end{array} \right\} \ldots\ldots\ldots(89);$$

and
$$C_l = C_r + \frac{\theta}{M} \frac{d_r l}{d\theta} \frac{d_l T}{d\theta} \quad \ldots\ldots\ldots\ldots(90),$$

or, by equation (88),
$$C_l = C_\tau - \frac{\theta}{M} \frac{\left(\frac{d_l T}{d\theta}\right)^2}{\frac{d_\theta T}{dl}},$$

or
$$C_l = C_\tau - \frac{\theta}{M} \frac{\left(\frac{d_\tau l}{d\theta}\right)^2}{\frac{d_\theta l}{dT}}.$$

Also
$$\frac{d_\phi \theta}{dl} = \frac{\theta}{M\bar{C}_l} \frac{d_l T}{d\theta} \quad \text{..............(91),}$$

and the condition that $d\phi$ is a complete differential, gives
$$\frac{d_\theta (MC_l)}{dl} = -\theta \frac{d^2_l T}{d\theta^2} \quad \text{..............(92).}$$

Lastly, taking T and l as independent variables, we have from equation (89),
$$dQ = MC_l \frac{d_l \theta}{dT} dT + \left(MC_l \frac{d_\tau \theta}{dl} - \theta \frac{d_l T}{d\theta}\right) dl,$$

or, by equation (90),
$$dQ = MC_l \frac{d_l \theta}{dT} dT + MC_\tau \frac{d_\tau \theta}{dl} dl \quad \text{..........(93),}$$

Hence
$$\frac{d_\phi T}{dl} = -\frac{C_\tau}{C_l} \frac{\frac{d_\tau \theta}{dl}}{\frac{d_l \theta}{dT}},$$

or
$$\frac{d_\phi T}{dl} = \frac{C_\tau}{C_l} \frac{d_\theta T}{dl} \quad \text{..............(94).}$$

Substituting for C_l, we obtain
$$\frac{d_\phi l}{dT} = \frac{d_\theta l}{dT} - \frac{\theta}{MC_\tau} \left(\frac{d_\tau l}{d\theta}\right)^2 \quad \text{..............(95).}$$

PART II.

CHANGE OF AGGREGATION.

101. Bodies are found in three different states of aggregation, known as the solid, the liquid, and the gaseous. Most substances are capable of existing in all three states. For example, water exists in the forms of ice, water, and steam. A few solids have not yet been melted, but the number of such bodies is found to diminish as improvements in the Arts and Sciences place higher temperatures at our disposal. Prior to 1877, the more perfect gases oxygen, hydrogen, and nitrogen, had resisted all attempts to reduce them to the liquid state; but in that year they were not only liquefied but solidified, by two independent experimenters, M. Cailletet and M. Raoul Pictet. It is therefore concluded that when we shall have a sufficient range of temperature and pressure at our command, it will be possible to make every substance take the three different forms of solid, liquid, and gaseous.

Let any quantity of any substance, as water, be contained in a cylinder fitted with an air-tight piston, so that the volume can be increased or diminished at pleasure; and suppose that, owing to the moderate size of the cylinder, the weight of its contents produces no sensible difference of pressure in any part of it. Then it is found that if the substance at any given temperature can exist in stable equilibrium in two different states in any proportion, it is possible, by altering the volume, to cause it to exist in stable equilibrium in these two states at the same temperature and pressure in all other proportions. We cannot therefore take θ and p for independent variables to define the state of the system when in equilibrium.

102. When a solid body is raised to a sufficiently high temperature, it begins to melt into a liquid. This change of state often takes place abruptly, as when ice is converted into water; but sometimes there are indications of an approaching change of state before that change actually occurs. For instance, glass, before reaching a state of perfect liquefaction, passes through a series of intermediate states in which it is soft, or viscous, and can readily be drawn out into very fine threads. But in all cases, when a body exists in the same vessel in a state of stable equilibrium at a given temperature and pressure, partly in the solid and partly in the liquid state, the proportion of the two parts may have any value we please. This temperature is called the melting point of the substance for the given pressure.

By carefully cooling a liquid in a clean vessel, it is found possible to reduce the temperature below the melting point without causing it to solidify. The liquid is then in a state of unstable equilibrium; the smallest shake or touch causing it to solidify with explosive violence.

Many solid bodies are constantly in a state of evaporation, or of transformation into the gaseous state. Camphor and ice are the best known examples of this. Such bodies, if not kept in well-stoppered bottles, gradually escape in the form of vapour. For example, large sheets of ice during a long dry frost become smaller and at last disappear. Very little is known of the conditions attending this phenomenon except from theoretical considerations.

APPLICATIONS OF CARNOT'S PRINCIPLE. 265

Again, if heat be applied to a liquid contained in a closed vessel, part of it will be converted into vapour or gas, and it is found that the pressure of the vapour when in stable equilibrium with the liquid, depends only on the temperature, increasing as the temperature rises. If the space above the liquid, instead of containing nothing but the vapour of the liquid, contain any quantity of air or any other gas not capable of chemical action on the liquid, it was found by Dalton that the quantity of vapour formed was very nearly as great as in the first case, but that the time required to reach the state of equilibrium was much longer.

103. As the transformation from the liquid to the gaseous state has been most studied experimentally, and as the phenomena are the same in all cases, we will describe the boiling of water by way of example.

When water is heated in an open vessel, the lowest layer gets hot first, and by its expansion becomes lighter than the colder water above and gradually rises, so that a gentle circulation is kept up whereby the whole of the water is warmed, though the lowest layer is always the hottest. As the temperature increases, the air, which is always absorbed to a small extent by cold water, is silently expelled, rising to the surface in small bubbles. At length, the lowest layer of water becomes so hot that in spite of the pressure of the atmosphere and the weight of the water above, a bubble of steam is formed which rapidly grows larger, and then rises, but is condensed by the colder water into which it ascends, the collapse producing the well-known noise of 'simmering' or 'singing.' In this way, the water is briskly agitated until it becomes hot enough throughout, and then the bubbles rise to the surface and the water is said to boil.

The temperature at which pure water boils depends chiefly on the pressure of the atmosphere, the greater the pressure the higher being the boiling temperature. But the water must be raised to a higher temperature than that at which the pressure of the steam is equal to that of the atmosphere, for the bubbles of steam have to overcome not only the pressure of the atmosphere, but the weight of the water itself. The dependence of the boiling point on the pressure may be clearly shown by connecting a closed glass vessel containing water and air with an air-pump by means of which the pressure of the air can be reduced at pleasure. After working the pump a short time, the water will be seen to enter into active ebullition. In this way it may be made to boil at any temperature between $0°$ C. and $100°$ C. On the other hand, by increasing the pressure of the air, the boiling point may be raised above $100°$ C.

A simpler experiment is due to Franklin. A little water is boiled in a flask over a gas flame until most of the air dissolved in it is expelled; and while the water is still boiling, the cask is securely corked and removed from the flame. Ebullition then ceases, but if cold water be poured over the vessel so as to condense the steam, the water again begins to boil and continues to do so for a considerable time.

104. In order to explain the indicator diagram of a substance, part of which is liquid and part vapour, we will suppose a gramme of water contained in a cylinder fitted with an air-tight piston. If the interior of the cylinder

APPLICATIONS OF CARNOT'S PRINCIPLE. 267

be large enough, the whole of the water will exist as steam satisfying approximately the relation $pv = R\theta$, where R is the constant 4,752,300. If, while the temperature is kept constant, we then force in the piston so as to diminish the capacity of the cylinder, the product pv will, at length, begin to decrease, and continue to decrease until the density of the vapour is exactly equal to that which is in stable equilibrium with water at the same temperature. The steam is then said to be saturated and the smallest increase of pressure is sufficient to cause some of it to be condensed into water. We may, if we are careful, convert all the steam into water without any appreciable change either in the temperature or the pressure.

The state of the substance before condensation begins is represented on the indicator diagram by the curve AB,

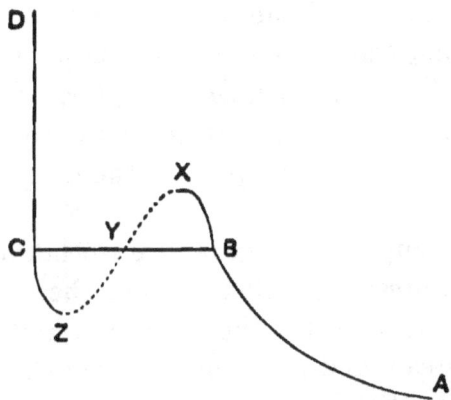

and during the change of state by the horizontal line BC. Also since water is very nearly incompressible, the remainder of the diagram will be the practically vertical line CD.

When the steam is in the state represented by the

point B, it is possible to keep the temperature constant and yet to increase the pressure without causing any steam to condense until the state represented by the point X is reached. In like manner, the pressure may be reduced considerably below that indicated by the point C before any water evaporates. But in both cases, the substance will be in an unstable condition, being in danger of explosive condensation on the curve BX, and of explosive evaporation on the curve CZ.

It has been suggested by Prof. J. Thomson that the curves BX and CZ may be continued into one another, as by the dotted line XYZ, so that the isothermal is only apparently and not really discontinuous. But experimental evidence of the existence of the curve XYZ is still wanting.

It will be observed that at any point in the curve XYZ, the pressure and the volume increase or diminish together, so that the state of the substance is then essentially unstable. In the present chapter, we restrict ourselves to stable conditions and reversible operations, and therefore the discontinuous part of the diagram is all we require.

In the accompanying figure, the isothermals for different temperatures are collected together. The dotted curve $B\,B'\,B''\ldots$ indicates the pressures and the volumes of saturated steam and is therefore called the 'steam line.' It is not an isothermal; for if it were, the different isothermal curves AB, $A'B'$, $A''B''$,... would all coincide with it, and for any given pressure and volume, the temperature of one gramme of steam might have an infinite number of values, which is plainly absurd. It is worthy of notice however, that a finite number of different temperatures are sometimes possible when the pressure and volume of a given mass of a substance are given. For example, at a

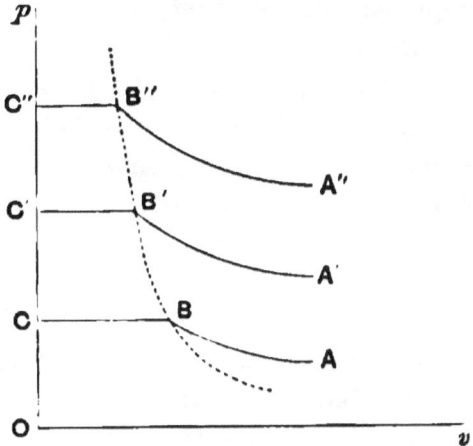

pressure of one atmo, a given quantity of water will have the same volume when its temperature is a little below 4° C. as when it is a little above.

The liquid state is practically represented by the axis of p, the volume of a given quantity of water at any ordinary temperature being negligible in comparison with the volume of the same quantity of saturated steam. The axis of p is therefore the 'water line.'

Along the steam line, there is evidently a relation between p and θ which may be written $p = f(\theta)$. A similar relation will hereafter be found to hold between p and θ when the solid and the liquid, or the solid and the gaseous states are in stable equilibrium together in the same vessel.

The steam line for water is represented approximately by either of the following empirical formulae, due to Rankine:

$$\left. \begin{array}{l} pv^{\frac{17}{16}} = \text{constant.} \\ v = (\theta - 233)^5 \times \text{constant.} \end{array} \right\}.$$

Since the density of saturated steam increases with the temperature, it follows that the steam and water lines continually approach one another, and we are naturally led to ask if they ever meet. This question has been answered by the experiments of Cagniard de la Tour and Dr Andrews, from which it appears that for each substance there is a certain temperature, known as its 'critical point,' above which the distinctions between the liquid and gaseous states disappear, whatever the pressure may be.

105. Let any sufficient quantity of any liquid and its saturated vapour be contained in a state of equilibrium in a cylinder fitted with an air-tight piston; and suppose that the cylinder is not so great but that the effect of gravity in creating differences of pressure within it may be neglected. Then if, by slowly drawing out the piston, an additional quantity of saturated vapour is formed in a reversible manner without altering either the temperature or the pressure, it is evident that the heat absorbed will be proportional to the quantity of saturated vapour thus produced, whether we adopt the caloric or the true theory of heat. The heat absorbed when an additional gramme of saturated vapour is formed in a reversible manner at any constant temperature θ and constant pressure p is called the Latent Heat of the vapour and is usually denoted by the symbol L. Since there is a relation between p and θ along the steam line, we may consider L to be a function of θ alone, or of p alone.

In order to explain how the subject was treated before the true theory of heat was established, we will first suppose heat to be a material substance, so that the total

APPLICATIONS OF CARNOT'S PRINCIPLE. 271

quantity of heat absorbed in any cyclical process is zero. Suppose then that the contents of the cylinder undergo the following cycle of operations.

(1) Let the piston be slowly drawn out until an additional gramme of saturated vapour is formed at constant temperature and pressure, whereby a quantity of heat L, is absorbed.

(2) Let the piston be drawn further out until the temperature falls from θ to $\theta - \tau$, where τ is indefinitely small, just sufficient heat being imparted to the cylinder or abstracted from it, to prevent the liquid from evaporating and the vapour from condensing.

As we wish to express the quantity of heat absorbed during this operation in a convenient form, we will represent by C' the heat that must be imparted to a gramme of saturated vapour to keep it constantly in the saturated state when it is slowly compressed until its temperature rises one degree: we will also suppose that, under similar conditions, C denotes the specific heat of the liquid, and H the thermal capacity of the contents of the cylinder in their original state. The heat absorbed in the operation may then be written

$$-(H - C + C')\tau.$$

The quantity C' is called the 'specific heat of the saturated vapour'; and since most liquids are nearly incompressible, it is evident that C is practically equal to C_p.

(3) Now let the piston be slowly pushed in until a gramme of vapour is condensed without altering the temperature or the pressure. The heat absorbed will be

$$-\left(L - \tau \frac{dL}{d\theta}\right).$$

(4) Lastly, let the contents of the cylinder be brought into their original state. The heat will be $H\tau$.

We have therefore

$$\frac{dL}{d\theta} + C - C' = 0 \quad\ldots\ldots\ldots\ldots\ldots(96).$$

Naturally, the first liquid experimented upon was water. It was ascertained by James Watt that the latent heat of steam diminished as the temperature increased, and he supposed his experiments to prove that the quantity of heat required to raise unit mass of water from the freezing point to any temperature θ at constant pressure and then to convert it into steam at that temperature and pressure, was independent of θ. This conclusion was known as Watt's law and was expressed by saying that, 'The sum of the free and latent heat is always constant'. In analytical language, this becomes

$$L + \int_{273}^{\theta} C_p d\theta = \text{constant.}$$

Differentiating, we obtain

$$\frac{dL}{d\theta} + C_p = 0,$$

and therefore, by equation (96),

$$C' = 0.$$

In accordance with this result, which was long taken to be correct, it was thought that when saturated steam expanded or was compressed in a vessel impermeable to heat, it would continue at the point of condensation. In other words, the steam line was considered to be an adiabatic curve.

In 1847, Regnault published his experiments on latent

APPLICATIONS OF CARNOT'S PRINCIPLE. 273

heat from which it appeared that Watt's law was not strictly correct, the value of $L + \int_{273}^{\theta} C_p d\theta$ increasing with the temperature. Regnault's result, which was expressed in calories, may be stated in the C. G. S. absolute system of units thus:

$$L + \int_{273}^{\theta} C_p d\theta = J(606{\cdot}5 + {\cdot}305\theta'),$$

where J is the number of ergs in a calorie, and $\theta' = \theta - 273$. Differentiating, we find

$$\frac{dL}{d\theta} + C_p = {\cdot}305J,$$

and then, by equation (96),

$$C' = {\cdot}305J.$$

It was therefore concluded that when saturated steam was compressed and consequently heated, it was necessary to supply heat to it from without to keep it in the state of saturation, and conversely, that it would have to part with heat when it expanded. From this it followed that when saturated steam was compressed in a vessel impermeable to heat, part of it would be condensed into water, but that if it was allowed to expand, it would be removed further and further from the saturated state; in other words, would become superheated.

106. The principles of the mechanical theory will now be applied to the reversible cycle described in the last article.

On an indicator diagram, the first and third operations will be represented by horizontal straight lines AB, CD, and the work done by the substance during the cycle, by the product of AB and PQ.

Now the ordinate OP represents the pressure p and OQ represents $p - \tau \frac{dp}{d\theta}$, where the value of $\frac{dp}{d\theta}$ is to be found by experiment from the properties of the steam

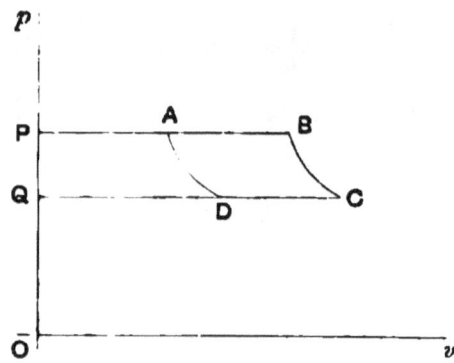

line. Also if s be the volume of one gramme of saturated steam, and σ that of one gramme of water, we shall have $AB = s - \sigma$. Thus if W be the work done by the substance during the cycle,

$$W = (s - \sigma) \tau \frac{dp}{d\theta}$$

$$= u\tau \frac{dp}{d\theta}, \text{ say.}$$

Hence the principle of energy gives

$$\frac{dL}{d\theta} + C - C' = u \frac{dp}{d\theta} \quad \ldots\ldots\ldots\ldots(97).$$

Again, the increase of entropy in the first operation is $\frac{L}{\theta}$ and in the third, $-\left\{\frac{L}{\theta} - \tau \frac{d}{d\theta}\left(\frac{L}{\theta}\right)\right\}$. In the second

APPLICATIONS OF CARNOT'S PRINCIPLE. 275

and fourth operations combined, it is $(C - C')\frac{\tau}{\theta}$. We have therefore

$$\frac{d}{d\theta}\left(\frac{L}{\theta}\right) + \frac{C - C'}{\theta} = 0,$$

that is,
$$\left.\begin{array}{c}\dfrac{dL}{d\theta} + C - C' = \dfrac{L}{\theta} \\[1ex] C' = \dfrac{dL}{d\theta} + C - \dfrac{L}{\theta}\end{array}\right\} \quad\ldots\ldots\ldots\ldots\ldots(98).$$

or

Combining equations (97) and (98), we get the important result

$$L = u\theta\frac{dp}{d\theta} \ldots\ldots\ldots\ldots\ldots\ldots(99).$$

107. Now in the case of water Regnault obtained the experimental results:

$$L + \int_{273}^{\theta} C_p d\theta = J(606{\cdot}5 + {\cdot}305\theta'),$$

$$C_p = J(1 + {\cdot}000\ 04\theta' + {\cdot}000\ 000\ 9\ \theta'^2).$$

Hence we find $\quad\dfrac{dL}{d\theta} + C_p = {\cdot}305J,$

and

$$L = J(606{\cdot}5 - {\cdot}695\theta' - {\cdot}000\ 0\ 2\ \theta'^2 - {\cdot}000\ 000\ 3\ \theta'^3),$$

so that, if we neglect the difference between C and C_p, and denote by c' the equivalent of C' in calories, we have

$$c' = {\cdot}305 - \frac{606{\cdot}5 - {\cdot}695\theta' - {\cdot}000\ 02\theta'^2 - {\cdot}000\ 000\ 3\ \theta'^3}{273 + \theta'}.$$

By means of this formula, the values of c' are easily calculated as in the table.

θ'	0	20	50	100	150	200
c'	$-1{\cdot}916$	$-1{\cdot}717$	$-1{\cdot}465$	$-1{\cdot}133$	$-{\cdot}879$	$-{\cdot}676$

Clausius finds that the simpler formula

$$L = J(607 - \cdot 708\theta')$$

may be substituted without serious error for the more complicated relation deduced by Regnault from his experiments. We then should have

$$c' = \cdot 305 - \frac{607 - \cdot 708\theta'}{273 + \theta'},$$

from which the values of c' may be calculated as before.

108. To find the difference between C and C_p, we take the equation

$$dQ = C_p d\theta - \theta \frac{d_p v}{d\theta} dp,$$

and suppose it to refer to one gramme of water in the liquid state, the relation between p and θ being the same as on the steam line. Thus we have

$$C = C_p - \theta \frac{d_p v}{d\theta} \frac{dp}{d\theta},$$

or
$$C_p - C = \theta \frac{d_p v}{d\theta} \frac{dp}{d\theta}.$$

Taking the temperature 100 C. for the purpose of numerical illustration, we then have, by experiment,

$$e = \cdot 000\,8,$$

or, since the volume of one gramme of water at 100° C. is practically one cubic centimetre,

$$\frac{d_p v}{d\theta} = \cdot 000\,8.$$

Also
$$\frac{dp}{d\theta} = 36{,}300.$$

APPLICATIONS OF CARNOT'S PRINCIPLE. 277

Thus at 100° C.,
$$C_p - C = 373 \times \cdot 000\ 8 \times 36{,}300.$$
But from Regnault's formula, we obtain at 100° C.
$$C_p = 1 \cdot 013 J.$$
Hence
$$C = J(1 \cdot 013 - \cdot 000\ 2\ 6).$$

It therefore appears that C and C_p are so nearly equal that no important error can have been introduced by taking them to be identical.

109. For Carbon Bisulphide (CS_2), Regnault finds
$$L + \int_{273}^{\theta} C_p d\theta = J(90 + \cdot 146\ 01\theta' - \cdot 000\ 412\ 3\ \theta'^2),$$
$$\int_{273}^{\theta} C_p d\theta = J(\cdot 235\ 23\theta' + \cdot 000\ 0815\theta'^2);$$
whence we obtain
$$L = J(90 - \cdot 089\ 22\theta' - \cdot 000\ 4938\theta'^2),$$
and
$$c' = \cdot 14601 - \cdot 000\ 8246\theta' - \frac{90 - \cdot 08922\theta' - \cdot 000\ 4938\theta'^2}{273 + \theta'}.$$

We then easily calculate the values of c' as below:

θ'	0	50	100	150
c'	$-\cdot 1837$	$-\cdot 16001$	$-\cdot 1406$	$-\cdot 1325$

For Chloroform ($CHCl_3$), Regnault finds
$$L + \int_{273}^{\theta} C_p d\theta = J(67 + \cdot 1375\theta'),$$
$$\int_{273}^{\theta} C_p d\theta = J(\cdot 232\ 35\theta' + \cdot 000\ 050\ 72\theta'^2);$$
and therefore
$$L = J(67 - \cdot 09485\theta' - \cdot 000\ 050\ 72\theta')$$
and
$$c' = \cdot 1375 - \frac{67 - \cdot 09485\theta' - \cdot 000\ 050\ 72\theta'}{273 + \theta'},$$

whence the following values of c' are deduced:

θ'	0	50	100	150
c'	$-\cdot1079$	$-\cdot0549$	$-\cdot0153$	$+\cdot0155$

For Ether ($C_4H_{10}O$),

$$L + \int_{273}^{\theta} C_p d\theta = J(94 + \cdot45\theta' - \cdot000\ 555\ 56\theta'^2),$$

$$\int_{273}^{\theta} C_p d\theta = J(529\theta' + \cdot000\ 29587\theta'^2);$$

and therefore
$$L = J(94 - \cdot079\theta' - \cdot000\ 85143\ \theta'^2),$$
and
$$c' = 45 - \cdot001\ 111\ 12\theta' - \frac{94 - \cdot079\theta' - \cdot000\ 85143\theta'^2}{273 + \theta'};$$

from which we find

θ'	0	50	100	150
c'	$+\cdot1057$	$+\cdot1222$	$+\cdot1309$	$+\cdot1344$

110. The fact that the specific heat of saturated steam is negative was discovered by Clausius and Rankine, independently, early in 1850. It shows that when saturated steam is compressed adiabatically it becomes superheated, and that when it expands it is partially condensed. In steam-engines the condensation is often prevented by means of a 'steam-jacket' surrounding the cylinder.

All the conclusions we have arrived at as to the specific heats of saturated vapours have been verified by the experiments of Hirn and Cazin. By employing a metal cylinder fitted with glass at the ends, the behaviour of the vapour was made visible to the eye. Bisulphide of carbon vapour and steam both formed

APPLICATIONS OF CARNOT'S PRINCIPLE. 279

a cloud during expansion but remained clear during compression. Ether, on the contrary, formed a cloud during compression and remained clear during expansion.

In the case of chloroform vapour, we have seen that the specific heat changes sign. Cazin calculates it to be zero at 123°·48 C. In accordance with this theoretical result, clouds were formed during expansion up to 123° C., but above 145° C. the vapour remained perfectly clear. Between 123° C. and 145° C. the conditions depended on the degree of expansion. With a small degree of expansion, there was no cloud; but with more expansion, a cloud appeared towards the end of the experiment, evidently depending on the temperature being reduced by expansion below 123°·48 C.

111. Experiments on latent heat or on the volume of saturated steam are somewhat uncertain, on account of the difficulty of ascertaining when the vapour is exactly in the saturated state and neither partially condensed nor superheated. But assuming Regnault's experimental results we have calculated the following table referring to saturated steam in which all the quantities are expressed in the C.G.S. system of absolute units.

The last three columns refer to one gramme of steam. To find s, we simply add σ to u, since $u \equiv s - \sigma$.

The values of $\frac{ps}{\theta}$ enable us to see the deviations of saturated steam from the state of perfect gas for which $\frac{ps}{\theta}$ is constant.

The values of $\frac{dp}{d\theta}$ are obtained from Clausius.

$\theta' = \theta - 273$	p	$\theta \frac{dp}{d\theta}$	Latent heat in ergs.	u	$\frac{ps}{\theta}$
0	6,134·40	120,021	25193,860000	209,910	4,717,000
5	8,713·52	166,696	25048,660000	150,270	4,710,000
10	12,222·1	229,373	24905,070000	108,580	4,689,000
15	16,934·9	312,054	24760,580000	79,347	4,667,000
20	23,192·0	420,073	24616,030000	58,599	4,638,000
25	31,405·5	557,431	24471,400000	43,900	4,627,000
30	42,071·3	732,128	24326,680000	33,227	4,614,000
35	55,779·1	952,166	24181,860000	25,397	4,600,000
40	73,220·8	1,225,550	24032,780000	19,610	4,588,000
45	95,203·3	1,561,610	23891,890000	15,300	4,581,000
50	122,661	1,973,680	23746,720000	12,032	4,570,000
55	156,661	2,471,100	23601,420000	9,551·4	4,562,000
60	198,416	3,068,540	23455,970000	7,644·0	4,555,000
65	249,294	3,781,990	23310,370000	6,163·5	4,547,000
70	310,830	4,626,140	23164,610000	5,007·1	4,538,000
75	384,734	5,618,310	23018,670000	4,097·1	4,533,000
80	472,904	6,777,180	22872,560000	3,374·9	4,521,000
85	577,437	8,121,400	22726,250000	2,798·3	4,515,000
90	700,645	9,684,360	22579,740000	2,331·6	4,502,000
95	845,070	11,479,300	22433,020000	1,954·2	4,486,000
100	1,013,510	13,530,400	22286,080000	1,647·1	4,478,000
105	1,208,760	15,853,400	22138,910000	1,396·5	4,469,000
110	1,434,080	18,499,200	21991,510000	1,188·8	4,455,000
115	1,692,840	21,473,100	21843,860000	1,017·3	4,443,000
120	1,988,720	24,816,300	21695,950000	874·3	4,430,000
125	2,325,580	28,551,700	21547,780000	754·7	4,416,000
130	2,707,510	32,716,400	21399,330000	654·1	4,401,000
135	3,138,850	37,335,900	21250,590000	569·1	4,387,000
140	3,624,140	42,444,700	21101,570000	497·1	4,372,000
145	4,168,130	48,072,400	20952,200000	435·8	4,358,000
150	4,775,810	54,249,500	20802,590000	383·5	4,343,000
155	5,452,370	61,020,000	20652,630000	338·5	4,330,000
160	6,203,240	68,402,600	20502,330000	299·7	4,310,000
165	7,033,950	76,436,000	20351,690000	266·3	4,300,000
170	7,950,270	85,156,200	20200,410000	237·2	4,280,000
175	8,958,140	94,595,200	20049,360000	211·9	4,260,000
180	10,063,600	104,766,000	19897,650000	189·9	4,250,000
185	11,272,900	115,725,000	19745,550000	170·6	4,230,000
190	12,592,500	127,496,000	19593,070000	153·7	4,220,000
195	14,028,600	140,094,000	19440,190000	138·8	4,200,000
200	15,588,000	153,565,000	19286,910000	125·6	4,180,000

APPLICATIONS OF CARNOT'S PRINCIPLE. 281

112. We will now obtain Clausius' expression for the energy and entropy of a substance of mass m, existing in a state of stable equilibrium, partly as liquid and partly as vapour.

Let us denote by U_0 and ϕ_0 the energy and entropy which the substance possesses when entirely in the liquid state at a given temperature θ_0 and a pressure equal to that of its saturated vapour at the same temperature. Also let U and ϕ be the energy and entropy when the temperature is θ and the mass of vapour x.

We may bring the substance from the first state to the second by the following reversible operations:—

(1) Let the temperature be gradually raised from θ_0 to θ without evaporating any of the liquid, the pressure being varied with the temperature in such a way that at every instant it is exactly equal to that of the saturated vapour.

The heat absorbed will be $m \int_{\theta_0}^{\theta} C d\theta$; the increase of entropy, $m \int_{\theta_0}^{\theta} \frac{C}{\theta} d\theta$; and the work done by the expansion of the liquid, $m \int_{\theta_0}^{\theta} p \frac{d\sigma}{d\theta} d\theta$; where C and σ have the same meanings as before.

(2) Then let the mass x of vapour be formed at constant temperature and pressure. The increase of entropy will be $\frac{xL}{\theta}$ and the increase of energy

$$x(L - pu).$$

Hence
$$\phi = \phi_0 + m \int_{\theta_0}^{\theta} \frac{C}{\theta} d\theta + \frac{xL}{\theta},$$

and $U = U_0 + m \int_{\theta_0}^{\theta} \left(C - p \frac{d\sigma}{d\theta} \right) d\theta + x(L - pu),$

or $U = U_0 + m \int_{\theta_0}^{\theta} \left(C - p \frac{d\sigma}{d\theta} \right) d\theta + xL \left(1 - \dfrac{p}{\theta \frac{dp}{d\theta}} \right)$ ...(100).

The equation $dQ = \theta d\phi$ then gives
$$dQ = mC d\theta + \theta d \left(\frac{xL}{\theta} \right) \quad \ldots \ldots \ldots (101).$$

The last result may also be obtained thus. The volume v of the substance at the temperature θ is
$$v = (m - x)\sigma + xs,$$
$$= m\sigma + xu.$$

Hence in a small reversible change of state, the work done on the substance is
$$dW = -p dv$$
$$= -mp \frac{d\sigma}{d\theta} d\theta - p d(xu).$$

The equation $dU = dQ + dW$ then gives
$$dQ = mC d\theta + d(xL) - xu dp$$
$$= mC d\theta + d(xL) - \frac{xL}{\theta} d\theta,$$
$$= mC d\theta + \theta d \left(\frac{xL}{\theta} \right), \text{ as before.}$$

113. If we suppose the substance, when partly liquid and partly saturated vapour, to expand adiabatically, we

APPLICATIONS OF CARNOT'S PRINCIPLE.

can easily calculate by means of the preceding formulae the change in the relative proportions of the liquid and gaseous portions of the substance, the change in volume, and the work done, in terms of the initial and final temperatures.

Since the entropy remains constant, we have, distinguishing the initial and final states by the suffixes (1) and (2),

$$m \int_{\theta_0}^{\theta_1} \frac{C}{\theta} d\theta + \frac{x_1 L_1}{\theta_1} = m \int_{\theta_0}^{\theta_2} \frac{C}{\theta} d\theta + \frac{x_2 L_2}{\theta_2},$$

or, since C is practically constant,

$$mC \log \frac{\theta_1}{\theta_0} + \frac{x_1 L_1}{\theta_1} = mC \log \frac{\theta_2}{\theta_0} + \frac{x_2 L_2}{\theta_2}.$$

Thus
$$x_2 = \frac{\theta_2}{L_2} \left(\frac{x_1 L_1}{\theta_1} - mC \log \frac{\theta_1}{\theta_2} \right) \ldots\ldots (102).$$

If v be the total volume of the substance,

$$v = (m - x)\sigma + xs$$
$$= m\sigma + xu$$
$$= m\sigma + \frac{xL}{\theta \frac{dp}{d\theta}}.$$

Hence
$$v_2 = m\sigma_2 + \frac{1}{\frac{dp}{d\theta_2}} \left(\frac{x_1 L_1}{\theta_1} - mC \log \frac{\theta_1}{\theta_2} \right) \ldots (103).$$

It has already been shown that the work, dW, done on a substance in an indefinitely small change of state is not generally a complete differential, and that, in consequence, the work W, done in a finite change of state, generally depends on the manner in which the change is effected. In the present case, however, we have the condition that the liquid and its saturated vapour are always in stable equilibrium together. The path being specified, the temperature may be considered as the only independent variable and the expression for dW becomes a complete differential. Thus

$$dW = -pdv = -pd(m\sigma) - pd(xu),$$

or, since the small quantity $m\sigma$ is sensibly constant,

$$dW = -pd(xu)$$
$$= -d(xup) + xu\frac{dp}{d\theta}d\theta.$$

Substituting from equation (99), this becomes

$$dW = -d(xup) + \frac{xL}{\theta}d\theta.$$

But, since ϕ is constant, we have from equation (100) or (101),

$$mCd\theta + \theta d\left(\frac{xL}{\theta}\right) = 0,$$

and therefore

$$\frac{xL}{\theta}d\theta = mCd\theta + \frac{d}{d\theta}(xL)d\theta.$$

Hence $$dW = -d(xup) + mCd\theta + \frac{d}{d\theta}(xL)d\theta.$$

Integrating and remembering that C is practically constant, we obtain

$$W = -x_2(u_2p_2 - L_2) + x_1(u_1p_1 - L_1) + mC(\theta_2 - \theta_1)$$
$$= -L_2x_2\left(\frac{p_2}{\theta_2\frac{dp}{d\theta_2}} - 1\right) + L_1x_1\left(\frac{p_1}{\theta_1\frac{dp}{d\theta_1}} - 1\right)$$
$$+ mC(\theta_2 - \theta_1)\ldots\ldots(104).$$

As an illustration of equations (102), (103), and (104),

the following table has been calculated by Clausius, work being reckoned in gramme-centimetres.

It is supposed that initially m grammes of saturated steam are contained at 150° C. in a cylinder impervious to heat, and that the piston is then slowly drawn out so that the temperature falls and the steam partially condenses. In our formulae, we must put $x_1 = m$.

$\theta_2 - 273$	150	125	100	75	50	25
$\dfrac{x_2}{m}$	1	·956	·911	·866	·821	·776
$\dfrac{m - x_2}{m}$	0	·044	·089	·134	·179	·224
$\dfrac{v_2}{v_1}$	1	1·88	3·90	9·23	25·7	88·7
$-\dfrac{W}{(980.868)\,m}$	0	1,130,000	2,320,000	3,590,000	4,930,000	6,370,000

114. When the liquid and gaseous states exist in stable equilibrium together, it is obvious to the most casual observer that there is a relation between the pressure and the temperature; but this is not so easy to see, without the assistance of theory, in the case of the solid and liquid states. In fact, the question whether the melting point depends on the pressure does not seem to have been asked before it was answered by Prof. J. Thomson in 1849, by the following train of reasoning.

For the sake of fixing the ideas, let the substance considered be water, which contracts in passing from the solid to the liquid state; and suppose the melting point independent of the pressure. Let a large quantity of it, partly in the liquid and partly in the solid state, at temperature θ and pressure p, be made to undergo the following cycle of reversible operations during which temperature is kept constantly equal to θ.

(1) Compress the mixture slowly until the pressure rises to $p + dp$, without causing all the water to be frozen. If our hypothesis be correct, this will be possible; for we have assumed that at a given temperature the liquid and the solid states may exist together at all pressures.

(2) Then by slowly abstracting heat, let a finite quantity of the water be frozen at constant temperature θ and constant pressure $p + dp$.

(3) Slowly reduce the pressure from $p + dp$ to p, without melting so much ice as was formed in the second operation.

(4) Lastly, let heat be imparted to the mixture so as to melt some of the ice at the constant temperature θ and constant pressure p, until the original state of the mixture is restored.

By drawing an indicator diagram, it will be seen that there is a gain of mechanical work in the cycle, which, by the principle of the equivalence of heat and work, must

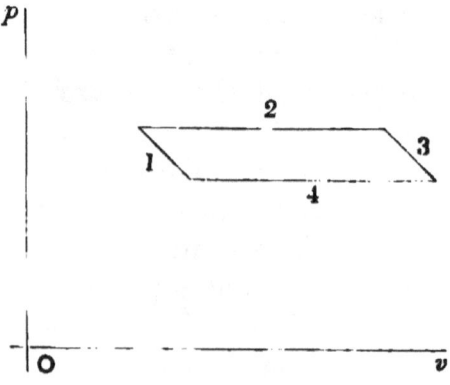

have been transformed out of the heat supplied to the substance. But this is contrary to Carnot's axiom, for all the heat gained or lost by conduction or radiation may have been obtained from or given to bodies whose temperature is uniform and constantly equal to θ.

We conclude therefore that the assumption that the melting point is independent of the pressure, must be incorrect, and we infer that there is a relation between p and θ when the solid and liquid forms are in stable equilibrium together. In like manner it may be shown that there is a relation between p and θ when the solid and gaseous states are in stable equilibrium together.

If we take two rectangular axes as axes of pressure and temperature, the relations between p and θ will be represented by three curved lines, of which that which refers to the solid and liquid states is called by Prof. J.

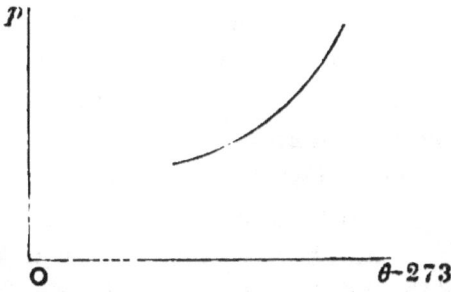

Thomson, the 'ice line'; that which refers to the solid and gaseous states, the 'hoar-frost line;' and that which refers to the liquid and gaseous states, the 'steam line.' The term 'steam line' is also used in a different sense, as we have already seen, as the name of the curve on the indicator diagram which represents the relation between the pressure and the volume of unit mass of saturated vapour.

115. The relation between the melting point and the pressure is given by an equation exactly similar to (99), viz.,

$$L = u\theta \frac{dp}{d\theta},$$

where L is the latent heat of fusion (of one gramme) in ergs, u the difference between the volumes of one gramme of the substance, when just at the melting point, first in the liquid and then in the solid state, and $\frac{dp}{d\theta}$ refers to the ice line.

Hence, since L is always positive, if the substance expand in melting, like most of the bodies which compose the crust of the earth, $\frac{dp}{d\theta}$ will be positive, or the melting point will be raised by increasing the pressure. In popular language, we may say that the greater pressure renders the increase of volume more difficult and that in consequence, a higher temperature is required for fusion than when the pressure is less.

The temperature of the earth increases rapidly as we descend, and at a moderate depth must be sufficient to melt every substance with which we are acquainted when the pressure to which they are subjected is only the same as that of the atmosphere. But it does not follow that the interior of the earth is in the liquid state, because the enormous pressure which must exist there may be more than sufficient to prevent liquefaction.

There are a few substances, like water and cast-iron; which contract in bulk when fusion takes place, so that u and $\frac{dp}{d\theta}$ will be negative. An increase of pressure will then lower the melting point, or assist the fusion.

APPLICATIONS OF CARNOT'S PRINCIPLE.

In the case of water at 0° C. and at the pressure of one atmo, we can easily find the value of $\frac{dp}{d\theta}$. For the latent heat is 79·25 calories, and the volumes of one gramme of water and of one gramme of ice are 1·000116 and 1·087 cubic centimetres, respectively. We therefore have

$$\frac{d\theta}{dp} = -\frac{273 \times \cdot 086884}{79 \cdot 25 \times 41{,}539{,}759 \cdot 8}.$$

If dp' be the pressure dp measured in atmos,

$$dp = 1{,}013{,}510\, dp',$$

and therefore

$$\frac{d\theta}{dp'} = 1{,}013{,}510\, \frac{d\theta}{dp}$$

$$= -\frac{273 \times \cdot 086884 \times 1{,}013{,}510}{79 \cdot 25 \times 41{,}539{,}759 \cdot 8}$$

$$= \cdot 0073025.$$

This theoretical result was verified experimentally by Sir W. Thomson in 1850 in a very accurate manner. He placed a quantity of water and lumps of clear ice in an Oersted press which was fitted with an ordinary air-gauge to show the pressure. In order to be able to measure small differences of temperature correctly, he constructed a thermometer filled with ether-sulphide and enclosed it in a larger glass tube hermetically sealed to protect it from the pressure. On screwing down the press, the temperature was at once seen to fall, but the thermometer returned to its original reading when the pressure was taken off. The results obtained are given below.

Pressure.	Fall of temperature of melting point.	
	Observed.	Calculated.
8·1 atmos	·059° C.	·059° C.
16·8 ,,	·129° C.	·1227° C.

The small difference between theory and observation when the pressure is 16·8 atmos may be due entirely to the fact that the theoretical equations refer only to indefinitely small changes of pressure.

The preceding conclusions with respect to ice have been applied to explain two phenomena which formerly occasioned much difficulty.

(1) When two pieces of ice at the melting point are pressed together, there will be a fall of temperature and some of the ice will be melted. For since at a temperature below the melting point, ice expands by heat at constant pressure, it follows from equation (70), that until the melting point is reached, an adiabatic increase of pressure will be attended by a rise of temperature. Hence, as may also be inferred from equation (102), when ice at the melting point is slightly compressed without loss or gain of heat, a small portion of it will be converted into water and there will be a very slight fall of temperature such that the water and ice may coexist at the higher pressure and lower temperature. The greater part of the water thus formed escapes, so that the two pieces of ice are left nearly dry. If the pressure be then diminished, the temperature will rise and the two pieces of ice will be frozen together, there being enough moisture left for this purpose.

This phenomenon is known as Regelation.

(2) It was noticed by Forbes in 1842 that a glacier descends along its bed with a motion like that of a very viscous fluid, such as tar. His observations were summed up in the words: 'A glacier is an imperfect fluid, or a viscous body, which is urged down slopes of a certain inclination by the mutual pressure of its parts.'

APPLICATIONS OF CARNOT'S PRINCIPLE.

This is easily explained by means of the principles which we have already laid down; for a glacier being a very porous mass of ice, the pressure will vary greatly from point to point; and the temperature being always about 0° C., the ice will melt at the places where the stress is most severe. The glacier will consequently be able to take new forms without exhibiting any visible rupture.

The behaviour of substances which expand during fusion, like wax and sulphur, was first examined experimentally by Bunsen by means of a very simple and ingenious apparatus. He took a glass tube, heated the ends, drew them out, and bent one of them round, as in the figure. The thick part B, of the tube was filled with mercury, the substance to be experimented upon introduced into the bent part C, and then both ends of the

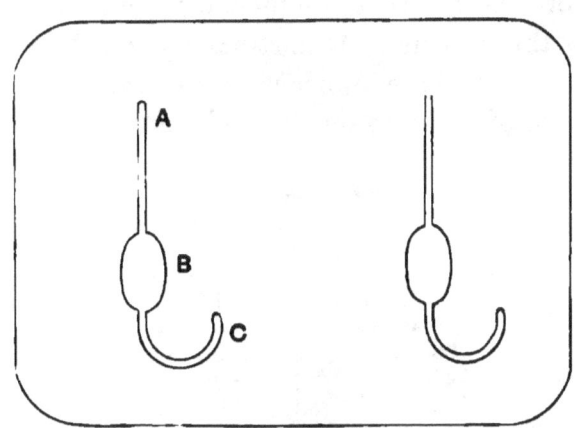

apparatus were hermetically sealed and the tube fastened to a board. The temperature of the bent part C could be varied at will by plunging it into water of a known temperature. On sinking the apparatus still deeper in the water, the mercury in B was heated and expanded, causing a great increase of pressure, the magnitude of which depended, of course, on the depth to which the part B was immersed in the water. It was capable of rising to above 100 atmos and was very accurately measured by the volume of the air in the fine tube A.

The method of making the experiment consisted in melting the substance in C and then, on allowing the water to cool, observing the temperature and the pressure at which solidification took place. In order to show simultaneously the behaviour of the substance under a pressure of about one atmo, another apparatus was constructed similar to the first, but with the end A open, and fastened on the same board.

The two substances examined by Bunsen were spermaceti and paraffin, and the results obtained, which are given below, showed that the melting point was raised by increasing the pressure. Unfortunately, we do not possess sufficient data to make an accurate theoretical calculation of the effects of pressure on the melting point.

SPERMACETI.

Pressure.	Melting point.
1 atmo	47·7° C.
29 atmos	48·3° C.
96 ,,	49·7° C.
141 ,,	50·5° C.
156 ,,	50·9° C.

PARAFFIN.

Pressure.	Melting point.
1 atmo	46·3° C.
85 atmos	48·9° C.
100 ,,	49·9° C.

APPLICATIONS OF CARNOT'S PRINCIPLE. 293

116. The latent heat of fusion depends on the temperature of fusion in the same way as the latent heat of evaporation depends on the temperature of evaporation; the relation between the latent heat of fusion and the temperature being of the same form as equation (98), viz.,

$$\frac{dL}{d\theta} + C - C' = \frac{L}{\theta},$$

where C and C' are the specific heats (in ergs) of the solid and liquid, respectively, when the pressure varies with the temperature so as to keep the solid on the point of melting and the liquid on the point of solidifying.

To find C and C' we have the equation

$$dQ = C_p d\theta - \theta \frac{d_p v}{d\theta} dp,$$

and therefore $\quad C \text{ or } C' = C_p - \theta \frac{d_p v}{d\theta} \frac{dp}{d\theta},$

C_p and $\frac{d_p v}{d\theta}$ referring, in the first case, to the solid state, in the other, to the liquid, and $\frac{dp}{d\theta}$ to the ice line in both cases.

If the substance considered be water at $0°$ C. and at the pressure of one atmo, the value of $\frac{dp}{d\theta}$ may be taken from Art. 115, thus:

$$C \text{ or } C' = C_p + \frac{79 \cdot 25 \times 41{,}539{,}759 \cdot 8}{\cdot 086884} \frac{d_p v}{d\theta}.$$

Hence if c, c', c_p be the equivalents of C, C', C_p, in calories,

$$c \text{ or } c' = c_p + 912 \frac{d_p v}{d\theta}.$$

Now the coefficient of cubical dilatation by heat at $0°$C. and at the pressure of one atmo, is $-\cdot 000\,061$ for water and ·000153 for ice. Hence, since the volume of one gramme of water in cubic centimetres is 1·000116 and the volume of one gramme of ice 1·087, we obtain

$$\frac{d_p v}{d\theta} = -·000061 \text{ for water,}$$

and

$$\frac{d_p v}{d\theta} = ·000166 \text{ for ice.}$$

Therefore
$$c = ·48 + ·1514 = ·6314,$$
$$c' = 1 - ·0556 = ·9444.$$

Consequently, if l be the latent heat of fusion of ice at 0° C., in calories, we obtain

$$\frac{dl}{d\theta} = \left(\frac{79·25}{273} + ·9444 - ·6314\right)$$
$$= ·290 + ·313$$
$$= ·603.$$

For the relation between the latent heat and the pressure, we have

$$\frac{dL}{dp} = \frac{dL}{d\theta}\frac{d\theta}{dp}$$
$$= -·603 \times ·2993$$
$$= -·1804779.$$

If dp' be the pressure dp expressed in atmos,

$$dp = 1{,}013{,}510 \times dp',$$

and therefore
$$\frac{dL}{dp'} = \frac{dL}{dp}\frac{dp}{dp'}$$
$$= 1{,}013{,}510 \frac{dL}{dp},$$

or
$$\frac{dl}{dp'} = \frac{1{,}013{,}510}{41{,}539{,}759·8}\frac{dL}{dp}$$
$$= -·0044.$$

APPLICATIONS OF CARNOT'S PRINCIPLE. 295

If dp'' be the pressure dp in grammes per square centimetre,
$$dp'' = 1033\cdot 279 \times dp',$$
and
$$\frac{dl}{dp''} = \frac{dl}{dp'}\frac{dp'}{dp''}$$
$$= -\cdot000\ 004\ 26.$$

117. The change from the solid to the gaseous state resembles the change from the solid to the liquid, or from the liquid to the gaseous state, and gives rise to equations exactly similar; but owing to the want of experimental data, they do not possess much interest. There is, however, one general point of great importance, which we will now explain.

Suppose a quantity of vapour at a high temperature contained in a vessel of constant volume and let the temperature be slowly reduced. After a time, the vapour will begin to condense into the liquid state, and as the fall of temperature goes on, the amount of vapour will become less and less. At length, the liquid will begin to freeze and the three different states of aggregation will be in stable equilibrium with one another. We see, then, that the three thermal lines, the steam line, the ice line, and the hoar-frost line, meet in a point which is consequently called the Triple Point.

The three thermal lines and the triple point X, for water, are shown in the accompanying diagram in which the abscissa of a point is proportional to $\theta - 273$, and the ordinate to p; OP representing a pressure of one atmo.

Since the ice line is nearly parallel to Op, the equation $\frac{d\theta}{dp'} = -\cdot 0073025$, which, by Art. 115, holds for the ice line at the point P, shows that the temperature of the triple point X lies between $0°C.$ and $·01°C.$

We shall find it convenient to distinguish the solid state by the suffix$_{(1)}$, the liquid by the suffix$_{(2)}$, and the

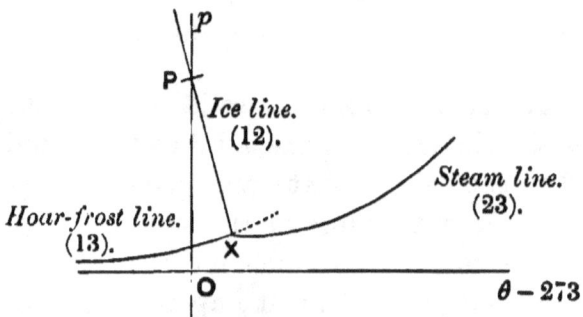

gaseous by the suffix$_{(3)}$. The steam line may then be distinguished by the suffix $_{(23)}$, the ice line by $_{(12)}$, and the hoar-frost line by $_{(13)}$. With this notation, we have, at the triple point X, but nowhere else,

$$L_{13} = L_{12} + L_{23},$$

and
$$u_{13} = u_{12} + u_{23}.$$

But if θ_0 be the temperature of the triple point X, the equations

$$L_{13} = u_{13}\,\theta_0\,\frac{dp}{d\theta_{13}}, \quad L_{23} = u_{23}\,\theta_0\,\frac{dp}{d\theta_{23}},$$

give us at X,

$$\frac{dp}{d\theta_{13}} - \frac{dp}{d\theta_{23}} = \frac{1}{\theta_0}\left(\frac{L_{13}}{u_{13}} - \frac{L_{23}}{u_{23}}\right).$$

Here, since at the triple point X, u_{12} is so small in comparison with u_{13} and u_{23} that we may take u_{13} and u_{23} to be equal, we have, at X,

$$\frac{dp}{d\theta_{13}} - \frac{dp}{d\theta_{23}} = \frac{L_{12}}{\theta_0 u_{23}}.$$

APPLICATIONS OF CARNOT'S PRINCIPLE. 297

Consequently, since L_{12} is very approximately
$$79\cdot245 \times 41{,}539{,}759\cdot8,$$
we have at X,
$$\frac{dp}{d\theta_{13}} - \frac{dp}{d\theta_{21}} = \frac{79\cdot245 \times 41{,}539{,}759\cdot8}{273 \times 209400}$$
$$= 57\cdot578.$$

If dp' be the pressure dp expressed in atmos,
$$dp = 1{,}013{,}510 \times dp',$$
and therefore at X
$$\frac{dp'}{d\theta_{13}} - \frac{dp'}{d\theta_{23}} = \cdot000\ 056\ 8.$$

If dp'' be the pressure dp in grammes per square centimetre,
$$dp = 980\cdot868 \times dp'',$$
and at X,
$$\frac{dp''}{d\theta_{13}} - \frac{dp''}{d\theta_{23}} = \cdot0587.$$

Now the value of $\dfrac{dp''}{d\theta_{23}}$ is $\cdot447$ at $0°$C. and $\cdot608$ at $5°$C. It may therefore be taken to be $\cdot448$ at the triple point. At this point, then, we shall have $\dfrac{dp''}{d\theta_{13}} = \cdot5067$.

PART III.

118. Many solid substances, such as the metal platinum, have the power of absorbing gases to an appreciable extent within their pores. The absorptive powers of charcoal are so great that the gases absorbed must often be in as dense a condition as when they are liquefied by enormous pressure and intense cold. Many liquids, too, possess the power of absorbing gases; and, as in the case of solid absorbents, there are a few exceptional instances in which the absorptive powers are very great; the most notable example being the absorption of ammonia by water. The increase of volume of a liquid due to the absorption of gases is generally small in comparison with the volume of the gases themselves: in the case of solid substances, the increase of volume is mostly imperceptible.

Again, if a piece of salt or a quantity of sulphuric acid be thrown into a liquid, it will generally be dissolved and the liquid will again become homogeneous. As more and more salt or acid is added, the solution will get stronger and stronger, until a certain state is reached depending on the pressure and temperature and then, however much salt or acid is thrown in, it will be unaffected by the liquid, which is then said to be 'saturated.' At any given pressure and temperature, there is therefore a maximum limit to the strength of the solution. On the other hand, if we begin with a quantity of salt or sulphuric acid, we may add as much water and, consequently, make the solution as weak as we please. Lastly, the vapour emitted by the solution is found to be of the same composition as the vapour of the liquid used in forming the solution, and its density, when in stable equilibrium with the solution, is found to depend on the temperature and on the strength of the solution.

Suppose now that a cylinder fitted with an air-tight piston, contains a system of uniform temperature θ, composed of a given absorbing mass M and a given mass m of some other substance, as a gas, a salt, a liquid or its vapour,

APPLICATIONS OF CARNOT'S PRINCIPLE. 299

and suppose that a mass x of this substance is absorbed by M. Also let the pressure p be such that the contents of the cylinder are in a state of stable equilibrium. Then if we put $\frac{x}{M} \equiv h$, we may obviously take any two of the three quantities θ, p, h, as independent variables to define the state of the system within the cylinder. On drawing out the piston, the unabsorbed mass $m - x$ will generally increase. If, at the same time, the temperature be kept constant, it will be necessary to impart a quantity of heat to the system, depending on (1) the absorbing part $M + x$, (2) the unabsorbed mass $m - x$. To distinguish between these two quantities, we will suppose the whole mass m to be originally absorbed, or take $x = m$.

Let this system be made to undergo the following cycle of reversible operations.

(1) Let the piston be slowly drawn out, at the same time imparting or abstracting sufficient heat to keep the temperature constant, until a small quantity of the substance absorbed is set free; and denote the heat imparted by dQ and the consequent increase of volume by dv.

(2) Let the system expand adiabatically until the temperature falls to $\theta - \tau$, where τ is indefinitely small.

(3) Let the piston be forced in at the constant temperature $\theta - \tau$ by such a distance that the original state may be restored by a second adiabatic process.

Representing the cycle on an indicator diagram by the small parallelogram $ABCD$, let the isothermal CD be continued to a point Q which represents the state of the system at the temperature $\theta - \tau$ when it first ceases to be homogeneous. Then, since we can always vary the second operation so as to make τ positive or negative at will, we may always suppose Q to lie to the left of the vertical line AP. The continuous curve CDQ will therefore always

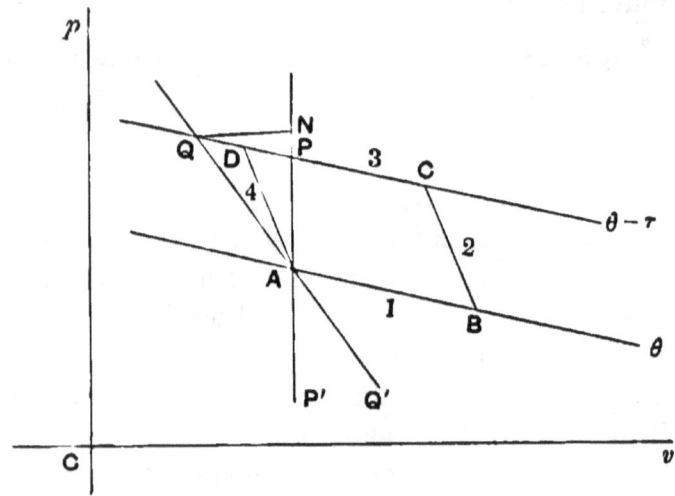

meet AP in some point P; and, since the projection of AB on the axis of v represents dv, the work done on the system in the cycle will be proportional to $AP\,dv$.

Again, when the pressure varies with the temperature in such a way as just to prevent the system from becoming heterogeneous, the coefficient of cubical dilatation by heat will generally be exceedingly small. The point Q and P will therefore practically coincide and the ratio of AN to AP may be taken to be unity, QN being perpendicular to AP. But the ordinate of A is proportional to p and that of Q to $p - \tau\dfrac{d_h p}{d\theta}$: hence AN, or AP, will be proportional to $-\tau\dfrac{d_h p}{d\theta}$, and the work done on the system during the cycle will be

$$-\tau\frac{d_h p}{d\theta}\,dv.$$

APPLICATIONS OF CARNOT'S PRINCIPLE.

Carnot's principle then gives

$$\frac{\tau \dfrac{d_h p}{d\theta} dv}{dQ} = \frac{\tau}{\theta},$$

or
$$dQ = \theta \frac{d_h p}{d\theta} dv \dots\dots\dots\dots (105).$$

If we put $dQ = Ldx$, and $dv = udx$, where dx is the mass of the substance set free in the first operation, we get

$$L = u\theta \frac{d_h p}{d\theta} \dots\dots\dots\dots (106).$$

When the substance absorbed is a gas, the volume of the small quantity dx will usually be enormous in comparison with the change of volume of the rest of the system, and may, without sensible error, be supposed equal to dv.

119. Again, suppose we have a pipe in which there is a plug of charcoal, and let a steady stream of gas or liquid be flowing slowly through it. Also let the temperature of every part of the plug be kept constant; the end A, at which the stream enters, being kept at a uniform constant temperature θ, and the other end B at a slightly different uniform constant temperature $\theta + d\theta$.

When one gramme of the gas or liquid passes through the plug, let the heat evolved at the end A of the plug be L, and the heat absorbed at the end B, $L + dL$. Also let the heat absorbed in the middle part of the plug exceed the quantity that would have been absorbed in the same time if the stream had not been flowing, by q.

Then dQ, the total quantity of heat absorbed by the unit mass of the stream in passing through the plug, will be
$$dQ = q + dL;$$
and if p be the pressure of the entering stream and $p + dp$ of the emergent stream, the increase of energy, dU, will be
$$dU = dQ - d(pv).$$
Also since the passage of the stream through the plug may be considered reversible, the increase of entropy of the unit mass in passing through the middle of the plug will be $\frac{q}{\theta}$, and $d\phi$, the total increase of entropy, will be
$$d\phi = -\frac{L}{\theta} + \frac{q}{\theta} + \frac{L + dL}{\theta + d\theta}$$
$$= \frac{q}{\theta} + d\left(\frac{L}{\theta}\right)$$
$$= \frac{q + dL}{\theta} - \frac{L}{\theta^2} d\theta$$
$$= \frac{dQ}{\theta} - \frac{L}{\theta^2} d\theta.$$

Therefore $\quad dU - \theta d\phi = -d(pv) + \dfrac{L}{\theta} d\theta \ \ldots\ldots$ (107).

Now if the substance which passes through the plug be water, which is practically incompressible, we shall have
$$d(pv) = 0,$$
and consequently $\quad dU - \theta d\phi = \dfrac{L}{\theta} d\theta \ \ldots\ldots\ldots$ (108).

Again, by Art. 112, we have for water
$$\left. \begin{array}{l} dU = C\, d\theta \\ d\phi = C\dfrac{d\theta}{\theta} \end{array} \right\},$$
so that $\quad dU - \theta d\phi = 0 \ \ldots\ldots\ldots\ldots$ (109).

Comparing equations (108) and (109), we obtain
$$L = 0.$$

If the substance which passes through the plug be air, or any of the perfect gases, $pv = R\theta$ and equation (107) becomes

$$dU - \theta d\phi = \left(-R + \frac{L}{\theta}\right) d\theta \quad \ldots \ldots (110).$$

But for a perfect gas,
$$\left. \begin{array}{l} dU = C_v d\theta \\ d\phi = C_p \dfrac{d\theta}{\theta} - \dfrac{v}{\theta} dp \end{array} \right\},$$

and therefore $\quad dU - \theta d\phi = -Rd\theta + vdp \quad \ldots \ldots (111).$
Hence, by equations (110) and (111),

$$L = v\theta \frac{dp}{d\theta} \quad \ldots \ldots \ldots \ldots \ldots (112).$$

Comparing equation (112) with equation (106), and remembering that when a gas is absorbed by charcoal, the change in the volume of the charcoal is practically zero, we see that $\dfrac{dp}{d\theta}$, which refers to the difference of pressure on the two sides of the plug, is equal to $\dfrac{d_h p}{d\theta}$. The meaning of this result, which has only been proved for an infinitesimal difference of temperature between the two ends of the plug, is that the mass of gas absorbed in each gramme of charcoal is the same in the hotter as in the colder part of the plug.

120. In 1858 an important formula was obtained by Kirchhoff which enables us to calculate the heat evolved on diluting a solution with water. The experiments of Thomsen on the heat evolved in the dilution of sulphuric acid and the observations of Babo on the pressure of the vapour emitted by dilute sulphuric acid, furnished results by means of which the formula was shown to be true.

Kirchhoff's formula may be easily deduced from equation (106) as follows. Let there be two cylinders fitted with air-tight pistons and joined by a narrow pipe which

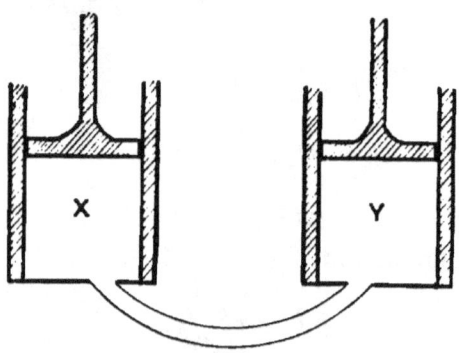

can be closed by a stop-cock; and suppose the solution contained in X and a small quantity of water dw in Y, both at the pressure of one atmo. Then let the same change of state be effected in the contents of the cylinders in two different ways during which the temperature of the whole system is kept constantly equal to θ.

(a)

Open the stopcock and let the water in Y be forced into X at the constant pressure of one atmo. Let the heat absorbed be denoted by dQ.

(b)

(1) Let the pressures in X and Y be reduced to p and P, respectively, so that the vapour is just on the point of forming in both cylinders. Let the increase of energy be called e_1.

APPLICATIONS OF CARNOT'S PRINCIPLE.

Since it is found by experiment that the pressure of the saturated vapour emitted by an aqueous solution is less than the pressure of saturated steam at the same temperature (it clearly could not be greater), it follows that P is greater than p.

(2) Let the water in Y be evaporated at the constant pressure P. The heat absorbed will be $u\theta \dfrac{dP}{d\theta} dw$, which may be written dQ_1, and the work done by the vapour on the piston is $Pudw$. Since the error committed by putting $Pu = R\theta$ is negligible at low temperatures in comparison with dQ_1, we obtain, very nearly, $dQ_1 = R\theta^2 \dfrac{1}{P}\dfrac{dP}{d\theta} dw$, and the increase of energy becomes $dQ_1 - R\theta dw$.

(3) Let the saturated steam in Y expand slowly at constant temperature until its pressure is equal to p. Since, at low temperatures, steam behaves somewhat approximately as a perfect gas, the increase of energy in this operation will be very small compared with dQ_1, and may be omitted altogether.

(4) Next open the stopcock and then force the vapour in Y slowly into X at constant temperature. Since the volume of the vapour is originally $\dfrac{R\theta}{p} dw$, the heat evolved during the operation will be $\dfrac{R\theta^2}{p} \dfrac{d_h p}{d\theta} dw$, which may be more briefly written dQ_2; and since the pressure never differs from p by more than an infinitesimal quantity, the work done on the system will be $R\theta dw$.

(5) Lastly, let the pressure be slowly increased to one atmo, and denote the corresponding increase of energy by e_2.

The total increase of energy in (b) will be

$$dQ_1 - dQ_2 + e_1 + e_2,$$

or, since e_1 is negative and both e_1 and e_2 very small in comparison with dQ,

$$dQ_1 - dQ_2.$$

Again, the increase of energy in (a) will practically be dQ, since the work done may be neglected. Hence

$$dQ = dQ_1 - dQ_2 \quad \ldots\ldots\ldots\ldots (113).$$

Substituting for dQ_1 and dQ_2 their values, this becomes

$$dQ = -R\theta^2 \left(\frac{1}{p}\frac{d_hp}{d\theta} - \frac{1}{P}\frac{dP}{d\theta} \right) dw$$

$$= -R\theta^2 \frac{d}{d\theta}\left(\log \frac{p}{P}\right) dw \quad \ldots\ldots (114).$$

121. We will now give some numerical illustrations of equation (106), depending on the following elementary proposition. Let there be two cylinders, as before, of which X contains a solution indefinitely near the point of saturation, and Y a small quantity dm of the substance absorbed, both at the same temperature θ and pressure p; and let the following change of state be brought about in the two following different ways at the constant temperature θ.

(a)

Open the stopcock and let the small mass dm be forced at constant pressure out of the cylinder Y into the solution X. Let the heat absorbed be denoted by dQ and the work done on the system by dW.

(b)

(1) Let the pressures on the pistons be slightly reduced until the stopcock can be opened without break-

APPLICATIONS OF CARNOT'S PRINCIPLE.

ing the equilibrium, and denote the increase of energy by e_1.

(2) Open the stopcock and then slowly force the mass dm into the solution, and let the heat absorbed be dQ' and the work done on the system dW'.

(3) Lastly, let the pressure on the piston of X be brought to its original value, the increase of energy being denoted by e_2.

Hence, by the principle of energy,
$$dW + dQ = dW' + dQ' + e_1 + e_2.$$
But dW, dW', e_1, and e_2, are small quantities of the second order. Thus
$$dQ = dQ'.$$
From this we see that when a small quantity of a substance is forced, at constant pressure and temperature, into an absorbing solution indefinitely near the point of saturation, the heat absorbed (or evolved) is the same as if the same substance had been forced at the same constant temperature into a large solution already saturated at the same pressure and temperature.

In the first place, consider the solution of carbonic acid by water. From experiment it appears that when the temperature remains the same, the amount of gas dissolved is very nearly proportional to the pressure, up to a pressure of 4 or 5 atmos; also that if a saturated solution be taken at the temperature 0°C. and at a pressure of one atmo, the quantity of gas dissolved is diminished in the ratio $\dfrac{1\cdot7207}{1\cdot7967}$ by raising the temperature to 1°C. without altering the pressure. In order, therefore, to preserve the strength of the solution constant when the temperature rises from 0°C.

to 1°C., the pressure must be increased in the ratio $\frac{1\cdot7967}{1\cdot7027} \equiv 1\cdot044284$. Thus at 0°C. and a pressure of one atmo, we have

$$\theta \frac{d_h p}{d\theta} = 273 \times \cdot 044284 \times 1013510.$$

The volume of one gramme of carbonic acid at 0°C. and a pressure of one atmo being $\frac{1000}{1\cdot9774}$ cubic centimetres, we see that when one gramme of carbonic acid at 0°C. and a pressure of one atmo is absorbed at this pressure and temperature by a large and nearly saturated aqueous solution of the same gas, the heat evolved, in calories, will be

$$\frac{273 \times 44\cdot284 \times 1013510}{1\cdot9774 \times 41539759\cdot8} \equiv 149\cdot17.$$

Again, the masses of equal volumes of water and carbonic acid, at 0°C. and a pressure of one atmo, are as 1000 to 1·9774, and therefore when one volume of carbonic acid is absorbed as before, the heat evolved is sufficient to raise the temperature of one volume of water by

$$\frac{149\cdot17 \times 1\cdot9774}{1000} \equiv \cdot 29$$

degrees Centigrade.

We now easily calculate that the heat evolved on forcing into solution at 0°C. and a pressure of one atmo, 1·7967 volumes of carbonic acid—the quantity required to saturate one volume of water at 0°C.—is sufficient to raise the temperature of one volume of water by 52°C.

At 0°C. and a pressure of one atmo, carbonic acid may be considered a perfect gas, and it will be seen that the result we have obtained relating to the absorption of one gramme of gas, is practically true at any smaller pressure;

for example, when a vessel of water absorbs carbonic acid from the air. In this case, the term 'pressure,' as we have used it, refers to the pressure of the carbonic acid alone, not to the sum of the pressures of the carbonic acid and the air.

Secondly, let us take the absorption of ammonia by water. In this case, at a pressure of one atmo, the quantity of gas absorbed by one gramme of water is ·875 gramme at 0°C. and ·833 gramme at 2°C., giving a difference of ·042 gramme. Also at 0°C., the quantity of ammonia absorbed by one gramme of water is ·875 gramme at a pressure of one atmo and ·906 gramme at 800 millimetres of mercury, or $\frac{20}{19}$ atmos. Thus the addition of $\frac{1}{19}$ atmo to the pressure increases the amount absorbed by ·031 gramme, and therefore, if we suppose that when the pressure does not differ much from one atmo, the quantity of gas absorbed is proportional to the pressure, the addition of $\frac{1}{19} \times \frac{42}{31}$ atmo to the pressure will increase the amount by ·042 gramme. Hence

$$\theta \frac{d_h p}{d\theta} = \frac{273}{2} \times \frac{42 \times 1013510}{19 \times 31}.$$

Now the volume of one gramme of ammonia at 0°C. and a pressure of one atmo, is $\frac{1000}{·7697}$ cubic centimetres. We see then that the heat evolved by the absorption of one gramme of ammonia, at 0°C. and a pressure of one atmo, by a large and nearly saturated solution of the same gas, is

$$\frac{273}{2} \times \frac{42 \times 1013510}{19 \times 31} \times \frac{1000}{·7697} \times \frac{1}{41539759·8} \equiv 309 \text{ calories}.$$

From this result we deduce that when ·875 gramme of ammonia—the quantity required to saturate one gramme of water at $0°$C. and a pressure of one atmo—is forced, as before, into a large saturated solution at this temperature and pressure, the heat evolved is 270 calories.

The experimental facts just made use of may be found in Storer's 'Dictionary of Solubilities,' or in Roscoe and Schorlemmer's 'Chemistry.'

Lastly, let us consider the solution of a salt in water. In several cases, this is necessarily an irreversible process. Thus if a vessel contain a saturated aqueous solution of chloride of barium or sulphate of magnesium together with an excess of the anhydrous salt, a rise of temperature will cause part of the free salt to be dissolved; but the process cannot be reversed, however slow the rise of temperature may be; for on reducing the temperature, the solution deposits a hydrate which falls peaceably among the free anhydrous salt. Still, when the rise (or fall) of temperature is slow enough, the system will be in equilibrium during the process, and all the properties of reversibility will be applicable.

In the case of the solution of common salt by water, at a pressure of one atmo, 100 parts by weight of water dissolve 36·7 parts of salt at 0° C. and 1·8 parts more at 100° C., the increase being practically uniform. Also a solution containing 75 grammes of water and 25 grammes of salt is of specific gravity 1·192, and therefore its volume is $\frac{100}{1\cdot192}$, or 83·89, cubic centimetres. Taking the specific gravity of the anhydrous salt as 2·15, the sum of the volumes of the water and salt separately would be $75 + \frac{25}{2\cdot15} \equiv 86\cdot62$ cubic centimetres. The total diminu-

APPLICATIONS OF CARNOT'S PRINCIPLE. 311

tion of volume is therefore a little greater than $\frac{1}{5}$ the volume of the salt added. Now it appears from experiment that when a small piece of salt is thrown into an unsaturated solution, the total diminution of volume is nearly independent of the strength of the solution. It is therefore always equal to $\frac{1}{5}$ the volume of the salt added.

Again, when one gramme of salt is thrown into a large and nearly saturated solution, the heat that must be imparted to keep the temperature constant is known to be 8·5 calories.

Hence, writing equation (105) in the more convenient form

$$dQ = \theta \frac{d_h p}{d\theta} dv = - \frac{\theta \dfrac{d_p h}{h\,d\theta}}{\dfrac{p}{h}\dfrac{d_\theta h}{dp}} p\,dv,$$

we find

$$\frac{p}{h}\frac{d_\theta h}{dp} = \frac{273 \times 100}{36\cdot 7} \times \frac{1\cdot 8}{100} \times \frac{1013510}{5 \times 2\cdot 15} \times \frac{1}{8\cdot 5 \times 41539759\cdot 8}$$
$$= \cdot 00354.$$

Putting $dp = \alpha p$, so that α is small, we get

$$\frac{1}{h} d_\theta h = \cdot 00354\, \alpha.$$

Hence, if H be the amount of salt dissolved in a saturated solution at $0°$ C. and a pressure of one atmo, the amount dissolved at the same temperature but at a pressure of $1 + \alpha$ atmos, where α is small, will be

$$H\,(1 + \cdot 00354\ \alpha).$$

122. The remainder of this chapter will be devoted to capillary phenomena. The systems studied will be supposed so small in mass that the mutual gravitation of their parts may be neglected, but so considerable in altitude that the attraction of the earth can no longer be left out of account.

If we take a glass tube of ordinary bore, and, after heating the middle, draw it out, we shall obtain a tube of small bore, known as a 'capillary tube.' On plunging this into water, it will be seen that the liquid rises in the tube above its proper level. On the other hand, if the tube be plunged into a trough of mercury, the mercury will stand at a lower level in the tube than elsewhere. This will explain the reluctance of mercury to enter a fine tube, which, it will be remembered, formed a difficulty to be overcome in making a thermometer.

When a liquid is elevated or depressed in a capillary tube whose bore is of any form and either uniform or variable, it will often be possible, by means of a slight touch or shake by which no appreciable amount of work is done, to cause the liquid to alter its level in the tube. Let us therefore suppose our system, consisting of liquid, capillary tube, and a considerable portion of the atmo-

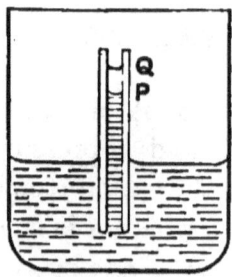

sphere, is enclosed in a large metallic vessel whose thickness is so great that its exterior surface always remains

APPLICATIONS OF CARNOT'S PRINCIPLE.

at the same uniform temperature θ, whatever actions may be going on inside. Then when the vessel and its contents are in a state of equilibrium at the uniform temperature θ, let a slight shake, by which no work is done, cause the liquid to change its level from P to Q, where PQ is either finite or infinitesimal. Afterwards let a second shake be given, and suppose, if possible, that the liquid in the tube returns to its original height P. Lastly, let the temperature throughout the vessel be made the same as at first without any further change in the height of the liquid in the tube.

Then the change in the energy of the vessel and its contents is zero, and therefore, since no work is done on it, the heat absorbed will also be zero. But by Carnot's principle, since the cycle is irreversible and the temperature of the exterior of the vessel always uniform and equal to θ, if Q be the heat absorbed by the vessel from without, $\frac{Q}{\theta}$ will be negative, so that Q must be negative. Hence the supposition that the liquid in the tube can be thus made to return from Q to P, is absurd.

From the preceding argument it will be seen that if a slight shake causes the liquid to rise from P to Q, it will be impossible, by a second slight shake, to cause it to descend below Q. Thence it may be inferred that there is some point O, above Q, to which successive slight shakes will cause the liquid to approach nearer and nearer. If the liquid stand originally above the point O, a slight shake will evidently cause it to descend; and when the level once coincides with O, no shaking will cause any further change. The point O is therefore a position of stable equilibrium at the temperature θ.

It might perhaps be supposed that at a given temperature θ and a given pressure of the atmosphere, there are several positions of stable equilibrium, O, O', O'',.... If this were the case, there would be a point A between two consecutive positions (O, O') of stable equilibrium such that if the liquid stood a little above A, a slight shake would make it rise still higher, and if it stood a little lower, a slight shake would cause it to descend. However it is easy to show, by experiment, that there is but one position of stable equilibrium; and in the rest of our discussions, we shall confine ourselves to the properties of this position.

123. Let a tube be constructed having its bore in its lower part very fine, but in the upper part considerable. Then if we plunge this tube into a liquid which rises in it, like water in a fine glass tube, it will be impossible for the liquid to rise into the wider part of the tube so long as that part is above the proper level of the liquid. For, if possible, let the liquid ascend into the wide part of the tube when that part of the tube is above the proper level

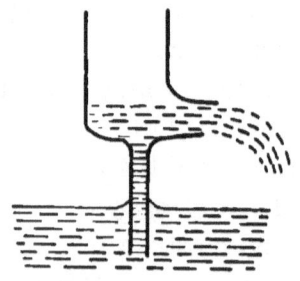

of the liquid, and let the bottom of the wide part be provided with a spout, like that of a pump, so that the

APPLICATIONS OF CARNOT'S PRINCIPLE.

liquid may run out, as in the figure. Then if a sufficient number of these self-acting pumps were employed, they would raise enough liquid to turn a small water-wheel which might be made to do mechanical work for us, in the usual way, by means of a pulley. This ideal contrivance might be in accordance with the principle of energy, for there might be enough heat absorbed to account for the mechanical work given out; but we can easily show that it cannot satisfy both the principle of energy and Carnot's principle. For let such an ideal system be constructed and let the whole of it, except the shaft and pulley by which the mechanical work is carried off, be enclosed in a large thick metallic vessel, the exterior surface of which is always at the same uniform temperature. Also let the capillary tubes be fitted with stopcocks so that the mechanism can be stopped or set in motion at will, without doing work or imparting or abstracting heat. Then when the vessel and its contents are in equilibrium at the uniform temperature θ, let the stopcocks be opened, and, after allowing the mechanism to work for a time, close them all again and let the system return to exactly the same state as before. Then since the change of energy is zero, if W be the work done on the vessel and its contents and Q the heat absorbed during the process, the principle of energy gives

$$W + Q = 0.$$

Also since the conception of entropy is applicable to the initial and final states, we have, by Carnot's principle,

$$\frac{Q}{\theta} < 0,$$

so that Q is negative.

We therefore have W positive. But, by hypothesis, there is a gain of mechanical work during the process, or W is negative. We therefore conclude that the original supposition in accordance with which the machine was to be constructed, is absurd.

We may vary the preceding proof by dispensing with the water-wheel and allowing the liquid raised by the self-acting pumps to fall back into the pool from which it was raised without doing mechanical work. We should then have the irreconcilable equations $Q = 0$ and $\frac{Q}{\theta} < 0$, from which we may draw the same conclusion as before.

Any ideal process, such as we have described, which is in contradiction to the principle of energy, or to Carnot's principle, or to both principles at once, is known as a 'perpetual motion.'

If we take a second tube the bore of which is every-

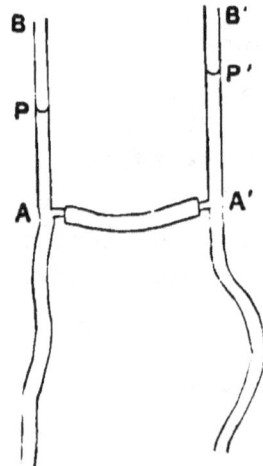

where fine and plunge it in a vessel of water or any other liquid which rises in the tube, then the height of the point P to which the liquid rises in the tube, will be independent of the form and bore of the lower part of the tube. For let AB be the upper part of the tube and let us take another capillary tube the upper part $A'B'$ of which is exactly similar and equal to AB; and suppose, if possible, that when this tube is plunged into the same vessel of liquid as the other with the part $A'B'$ placed parallel to and on the same level as AB, the point P', to which the liquid rises in $A'B'$, is not in the same horizontal plane as P. Let there be a stopcock at A and another at A', which can be turned without doing work or imparting or abstracting heat, and let these stopcocks be so constructed that they afford the means of closing the downward communication of the capillary tubes AB, $A'B'$, and of causing these tubes to connect with one another by means of a tube AA' of any kind. Lastly, let the whole system be enclosed in a metallic vessel, as before, so that heat can only be absorbed or evolved at the uniform constant temperature θ. Then when the system is in equilibrium at the uniform temperature θ with the tubes AB, $A'B'$ communicating downwards, let the stopcocks be turned until the tubes communicate only with one another, through the pipe AA'. This will evidently cause the liquid to assume the same level in AB as in $A'B'$, and there will therefore be a flow through the pipe AA', which may either be used as a source of mechanical work, or be allowed to expend itself wholly in friction. After a sufficient time, let the cocks be turned back so that the system returns exactly to its initial state. Then we have the contradictory equations

$W + Q = 0$, $W < 0$, $\dfrac{Q}{\theta} < 0$, or else $Q = 0$ and $\dfrac{Q}{\theta} < 0$. From either set of equations we conclude that the points P and P' must be on the same level.

Again, if we take a tube the bore of which is everywhere fine, except in the middle (as in the figure), and then plunge it deep enough in a liquid which it depresses

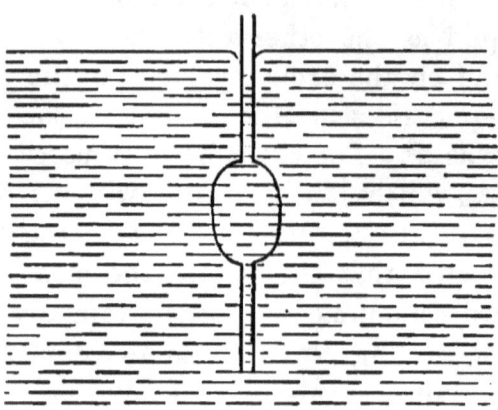

with the wide part entirely below the surface, it is clear that the liquid will fill the wide part of the tube and rise into the capillary part above it. It may also be shown that the depth of the surface of the liquid column in the capillary part of the tube is unaffected by the form of the lower part of the tube.

It will now be seen that the force by which the liquid is caused to rise, or is depressed, in the capillary tube, must be sought for at the surface of the liquid column in the tube. Also since it appears from experiment that the action of a solid on a liquid is insensible at measurable distances, we are forced to conclude that the surface of the liquid column in the tube must be in a state of tension and merely held to the tube at its circumference. From this inference we may anticipate the form

APPLICATIONS OF CARNOT'S PRINCIPLE. 319

of the surface, which, in the case of water in a glass tube,

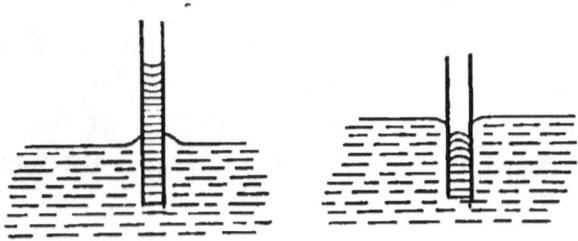

is concave upwards, but in the case of mercury in a glass tube, convex upwards.

124. Let us take a capillary tube whose bore is uniform and in the form of a small circular cylinder of diameter d, and let this tube be immersed vertically in a liquid which rises so high or is depressed so low, that the curvature of the top of the liquid column in the tube may be neglected in comparison. Also let the system be enclosed in a large metallic vessel so that heat can only be absorbed or evolved at the constant uniform temperature θ. Then if we assume that the action of the tube on the liquid extends only to an exceeding small distance from their common surface, it is obvious that when the system is in a state of equilibrium, stable or unstable, with the liquid in the tube at a height z above its proper level, with the temperature uniform throughout the vessel and equal to θ, and with the practically uniform pressure of the air contained within the vessel equal to p, we shall have

$$\left.\begin{array}{l} U = U_0 + \pi d z F_1(\theta, p) + \pi d F_2(\theta, p) \\ \phi = \phi_0 + \pi d z f_1(\theta, p) + \pi d f_2(\theta, p) \end{array}\right\} \ldots\ldots(115),$$

where U_0 and ϕ_0 are independent of z, and the four quantities (F_1, F_2, f_1, f_2) depend only on the natures and physical states of the tube and liquid so that they are independent of both z and d.

Now if a slight shake be given to the vessel, by which the liquid is caused to rise in the tube to the height $z + dz$, we shall have, if ρ be the density of the liquid and g the attraction of the earth in dynes on a mass of one gramme,

$$dU = -\frac{\pi}{4} d^2 g\rho z\, dz + dQ,$$

and $$dQ - \theta d\phi < 0.$$

Therefore $$dU - \theta d\phi + \frac{\pi}{4} d^2 g\rho z\, dz < 0,$$

so that, if the temperature be allowed to regain its former uniform value, the expression

$$U - \theta\phi + \frac{\pi}{8} d^2 g\rho z^2 \quad\ldots\ldots\ldots\ldots (116)$$

will have decreased during the operation.

Hence, since it is obvious that the expression must have a minimum value, when that value is reached the system will be in a state of stable equilibrium. To find the position of stable equilibrium at temperature θ we have therefore only to take the variation of the expression (116) on the supposition that θ is constant, and then equate the variation to zero.

This result may also be obtained thus. When the system is in a state of stable equilibrium at the uniform temperature θ, let $z + dz$ be the height of the liquid in the tube above its proper level; and suppose that when the liquid is in a state of equilibrium at the uniform temperature θ with the liquid only at the height z above its proper level, the vessel is carefully lifted up from the ground, or let down a pit, so that g slightly changes until the equilibrium becomes stable. Then let the vessel be brought to its original position in such a way that the liquid is caused to rise in the tube in a reversible manner to the height $z + dz$. Then since g only changes by an infinitesimal amount during the operation, the work done on the vessel and its contents will be $-\frac{\pi}{4} d^2 g\rho z\, dz$, and therefore

$$dU = -\frac{\pi}{4} d^2 g\rho z\, dz + dQ'.$$

Also, since the operation is reversible,

$$dQ' - \theta d\phi = 0.$$

Hence, as before, if θ remain constant and the equilibrium be stable,

$$d\left(U - \theta\phi + \frac{\pi}{8} d^2 g\rho z^2\right) = 0.$$

Thus if H be the value of z in the state of stable equilibrium at the temperature θ,

$$\pi d(F_1 - \theta f_1) + \frac{\pi}{4} d^2 g\rho H = 0,$$

or
$$4(F_1 - \theta f_1) + dg\rho H = 0 \quad \ldots\ldots\ldots (117).$$

Consequently, since F_1 and f_1 depend only on the natures and physical states of the tube and liquid, if we have tubes of the same substance in the same physical state but of different uniform bores, immersed in the same kind of liquid, dH will have the same value for all.

The force by which the liquid is raised or depressed in the capillary tube, reckoned positive when it acts upwards, is
$$\frac{\pi}{4} d^2 g\rho H,$$

or, by equation (117),
$$-\pi d(F_1 - \theta f_1),$$
which is equal to a force of $-(F_1 - \theta f_1)$ per centimetre of the circumference of the bore of the tube.

If two plates of the same nature as the tube just described, be placed parallel to one another at a distance d, and immersed vertically in the same kind of fluid as the tube; then if we consider only so much of the system as lies between two ideal vertical planes at right angles to the plates and at a distance of one centimetre apart, we shall have, when the system is in a state of equilibrium at the uniform temperature θ with the liquid standing between the plates at a height z above its proper level,
$$\left.\begin{array}{l} U = U_0' + 2zF_1(\theta, p) + 2F_2(\theta, p) \\ \phi = \phi_0' + 2zf_1(\theta, p) + 2f_2(\theta, p) \end{array}\right\} \ldots\ldots(118),$$
where U_0' and ϕ_0' are independent of z and d, and the functions (F_1, F_2, f_1, f_2) are the same as before. Also when the equilibrium is stable,
$$dU - \theta d\phi + dg\rho z dz = 0;$$
so that if h be the value of z in that case,
$$2(F_1 - \theta f_1) + dg\rho h = 0 \ldots\ldots\ldots\ldots(119).$$
Comparing equations (117) and (119), we see that
$$H = 2h,$$
or that the liquid is raised or depressed between the two plates exactly half as much as in the tube.

125. The earliest application of thermodynamics to capillary phenomena was made by Sir W. Thomson in 1870.

APPLICATIONS OF CARNOT'S PRINCIPLE.

If a fine glass tube A, open at both ends, be plunged with its lower extremity in a sheet of water, the water will be seen at once to ascend the tube, and it is obvious that in the state of stable equilibrium, the equilibrium will not be disturbed by closing the lower end of the tube. Hence we infer that if a second tube B, exactly similar

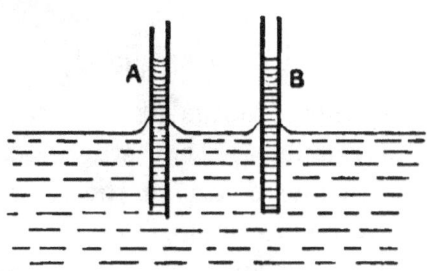

and equal to A, but with its lower end closed, be held parallel to and on the same level, as A, equilibrium will not be completely established, if there is any vapour in the air, until the water is at the same level in B as in A. The tube B must therefore have the power of condensing the aqueous vapour of the atmosphere. As this process goes on, the quantity of water inside the tube B will increase and the density of the vapour on its top become less and less, until at last the state of equilibrium is attained.

If p be the pressure and ρ the density of the vapour in the air just over the level sheet of water, and p' the pressure of the vapour at the height h of the water column, we have

$$p' = p - g\rho h,$$

since the variations of the density in the height h are small in comparison with ρ.

Hence
$$p' = p\left(1 - \frac{g\rho h}{p}\right).$$

Now if the temperature be $10°$ C., $p = 12{,}222$ and
$$\rho = \frac{1}{108{,}581}.$$

Thus, since we may take $g = 981$, we obtain, at $10°$ C.,
$$p' = p\left(1 - \frac{h}{1{,}352{,}800}\right).$$

Also in a tube whose radius is the thousandth of a millimetre, it is calculated that water would rise to a height of about 13 metres above the plane level. Consequently the equilibrium pressure of the aqueous vapour in the tube will be less than just over the plane surface of the water by about the thousandth part.

Sir W. Thomson thinks it probable that the moisture which vegetable substances, such as flannel, cotton, etc., acquire from the air at temperatures far above the 'dew-point'—that is, far above the temperature at which, in a given state of the air, vapour begins to be deposited on plane surfaces—may be accounted for by the condensation of vapour in the minute cells of the substance.

This article will also help us to understand the fact, which we have already noticed, that when the bulb of a thermometer is wrapped with flannel and then breathed upon, the temperature of the bulb rises so much above that of the breath.

CHAPTER V.

THE THERMODYNAMIC POTENTIAL.

126. The object of the present chapter is to find a simple method, corresponding to the method of the Potential in Statics, which shall determine whether the equilibrium of a system is stable or unstable, and give the properties of stable equilibrium.

The first attempt of this kind was made by Berthelot. The rule which he proposed under the name of the 'principle of maximum work' was very simple.
Thus if a system be in a state of equilibrium, the temperature will be uniform throughout. If now, without any interference from other systems, any disturbance or chemical reaction takes place, the temperature will be rendered variable but ultimately it will again become uniform. Berthelot then supposes that the final temperature is higher than the first, and that, consequently, if the temperature fall to its original value by conduction or radiation, the system must give out a positive quantity of heat. This assumption may be more simply expressed by saying that every chemical reaction tends to make the

system assume that state in the production of which it liberates most heat. Or again, since the system is not at liberty to receive or lose energy except in the form of heat, we may say that the energy of the system tends to a minimum value.

Berthelot's rule can be easily applied and its results are often in satisfactory agreement with experiment. There are, however, a few common cases in which it altogether fails. For example, if salt and snow be mixed together, there is a considerable fall of temperature, and if the original temperature be restored, there will be an absorption of heat and an increase of energy.

A correct test of stability is afforded by the theory of entropy. We have already seen that when any material system is prevented from receiving or losing heat, its entropy is constantly increasing except when the system is in a state of stable or unstable equilibrium. The equilibrium will therefore be stable when the entropy of the system is a maximum. For instance, if the system be in equilibrium in a state P for which the entropy is a maximum, it will be impossible for the system to pass from the state P to any neighbouring state Q, because this would require the entropy to be greater in the state Q than in the state P, since the path from P to Q can never be strictly reversible or non-frictional.

The method of entropy is not of much practical value, on account of the difficulty of making the calculations. A different test of stability is therefore used which is much simpler in application, because it supposes the temperature constant. The properties of the functions which are employed for this purpose appear to have been first investigated by Massieu in 1869, but he did not examine whether they furnished a test of stability. This was done independently by Prof. W. Gibbs in 1875 and by Dr Helmholtz in 1882. It is worthy of notice that the earlier Thermodynamicists have had no part in discovering the method.

For the sake of simplicity, we shall suppose every part of the bodies or systems which we consider to be readily permeable to heat, so that, in a state of equilibrium, the temperature will be uniform throughout. When such a body or system undergoes a small reversible modification, we have

$$\left.\begin{array}{l} dU = dQ + dW \\ dQ = \theta d\phi \end{array}\right\},$$

and therefore $dW = dU - \theta d\phi$.

Hence, if the temperature remain constant during the process,

$$dW = d(U - \theta\phi).$$

If we put $\mathcal{F}$ for $U - \theta\phi$, the preceding equation becomes

$$dW = d\mathcal{F}.$$

We see then that if a body undergoes a reversible operation during which the temperature is kept constant, the function $\mathcal{F}$, which depends only on the state of the body, will be increased by the amount of work done **on** the body, or decreased by the amount of work done **by** the body. For this reason, the function $\mathcal{F}$ is called by Helmholtz the 'free energy,' being that part of the energy which is convertible into mechanical work at the constant temperature θ. In like manner, $\theta\phi$ is called the 'bound energy,' because it is not convertible into work. No further reference will be made to this nomenclature, and it should be noticed that it has no connection whatever, beyond a similarity of sound, with our conceptions of free and bound etherial energy; the names of which have been suggested to us by reading Lodge's 'Modern views of Electricity.'

When the only external forces to which the body is subjected consist of a uniform and constant normal pressure, p, on the surface, $dW = -pdv$, so that if we write Φ for $\mathcal{F} + pv$, or $U - \theta\phi + pv$, the preceding result becomes

$$d\Phi = 0.$$

If, under these conditions, v be also constant, then

$$d\mathcal{F} = 0.$$

Let us now suppose that the system is in equilibrium in any state A at any uniform temperature θ. Then the equilibrium in the state A will be unstable if a slight shake or touch, by which no perceptible change is made in the system, causes the equilibrium to be broken in consequence of which the system rushes into some other state P. Hence clearly the equilibrium in the state A will be stable if every spontaneous change of state, like AP, is impossible.

In order to express this condition mathematically, let us suppose the system enclosed in a vessel which is so rigid that no mechanical work can be obtained from it, and so slow a conductor of heat, especially at its inner surface, that the temperature of its exterior surface is always uniform and equal to θ, whatever processes may be going on inside. Let us further suppose the vessel to be such that the same external forces act on the system when it is inside the vessel as when the vessel is not employed. Also let the whole of the vessel as well as its contents be originally at the uniform temperature θ, and after the spontaneous change of state from A to P, let us wait till the vessel and its contents are of the same uniform temperature as before, and call the new state of the system within the vessel B.

Then since there is no change in the state of the containing vessel, and since the total quantity of heat absorbed by it during the interval is zero, the heat absorbed by the system within the vessel will be equal to the heat absorbed by the vessel at its exterior surface. Hence if ΔQ be the heat absorbed by the system during the change of state AB, the magnitude of which may be finite, and $\Delta\phi$ the increase of entropy, we shall have

$$\int_A^B \frac{dQ}{\theta} < \Delta\phi,$$

or
$$\Delta Q < \theta\Delta\phi.$$

But if ΔU be the increase of energy of the system and ΔW the work done on it, the principle of energy gives

$$\Delta U = \Delta W + \Delta Q.$$

Hence $\qquad \Delta U - \Delta W < \theta\Delta\phi,$

or $\qquad \Delta U - \theta\Delta\phi < \Delta W.$

Thus since θ has the same values in both of the states A, B,
$$\Delta(U - \theta\phi) < \Delta W,$$

that is, $\qquad \Delta\mathcal{F} - \Delta W < 0.$

Now if the external conditions be such that ΔW is a complete differential, the value of $\Delta\mathcal{F} - \Delta W$ will depend only on the states A and B, and we shall be able to draw some remarkable conclusions. This will happen in two very simple and important cases to which we shall restrict ourselves for the future in discussing questions of stability. In both of these cases, the external force is a uniform normal pressure on the surface: in the first case, the volume is also constant, and in the second, the surface pressure is constant as well as uniform.

When the volume is constant, $\Delta W = 0$, and therefore $\Delta \mathcal{F}$ must always be negative. When the pressure is constant, $\Delta W = -p\Delta v = -\Delta(pv)$, and therefore $\Delta(\mathcal{F} + pv)$, or $\Delta \Phi$, must always be negative. Hence when the function $\mathcal{F}$, or the function Φ, attains a minimum value corresponding to any *given* temperature, the change of state from A to B will cease to be possible, and consequently also, the spontaneous change from A to P. In other words, A will be a state of stable equilibrium.

This may also be shown in a different way. For it will be possible, by violating the conditions of volume or pressure, to bring the system from A to B by a reversible operation during which the temperature is constantly uniform and equal to θ. If in this case, $\Delta Q'$ be the amount of heat absorbed, we shall have

$$\Delta Q' = \theta \Delta \phi.$$

Hence, when the volume is kept constant,

$$\begin{aligned}\Delta \mathcal{F} &= \Delta U - \theta \Delta \phi \\ &= \Delta Q - \theta \Delta \phi \\ &= \Delta Q - \Delta Q',\end{aligned}$$

and when the pressure is kept constant,

$$\Delta \Phi = \Delta Q - \Delta Q'.$$

Now when A, B are states of unstable equilibrium near together, the amount of work done on the system during the reversible path will clearly differ by a finite quantity from the work actually done when the system passes of itself from A to B under the given conditions of volume or pressure. The heat absorbed will therefore differ by a finite quantity in the two cases and the variations of $\mathcal{F}$ and Φ will consequently be finite. But when A, B are states of stable equilibrium, the difference in the work done or in the heat absorbed, will be a small quantity of the second order, which may be neglected. A state of stable equilibrium will therefore make $d\mathcal{F} = 0$ when the volume is constant, and $d\Phi = 0$ when the pressure is constant, *the temperature being supposed constant in effecting the differentiation.*

Since U and ϕ both contain an arbitrary constant, the functions $\mathcal{F}$ and Φ will each contain an arbitrary term of the form $\alpha + \beta\theta$, where α and β are constants. But this does not affect the condition of stability, because the temperature remains constant in taking the variations of $\mathcal{F}$ and Φ.

On account of the important properties just obtained, which include the 'energy test' of stability in advanced abstract dynamics, the function $\mathcal{F}$ is called by Duhem the 'Thermodynamic Potential at constant volume,' and Φ the 'Thermodynamic Potential at constant pressure.'

127. To obtain the formulae of M. Massieu, let us suppose the state of the substance to be completely defined in equilibrium by the temperature and volume, so that θ and v may be taken as independent variables. Then for a change from the state (θ, v) to the state $(\theta + d\theta, v + dv)$, we have

$$d\mathcal{F} = dU - \theta d\phi - \phi d\theta.$$

This result will not be true if θ and v are not sufficient to define completely the state of the system; for if the change of ϕ corresponding to $d\theta$ and dv were finite, the increment of $\theta\phi$ would not be $\theta\Delta\phi + \phi d\theta$. If the pressure on the surface during the change of state be uniform and equal to p, we have

$$dU = dQ - pdv,$$

and therefore $\quad dF = dQ - \theta d\phi - \phi d\theta - pdv.$

Now the preceding process will not generally be reversible, but if the equilibrium be stable, the same small change of state may be brought about at constant temperature θ by a reversible process in which the work done on the system only differs from $-pdv$ by a small quantity of the second order. We shall therefore have $dQ = \theta d\phi$,

and consequently, $\quad dF = -\phi d\theta - pdv,$

or
$$\left.\begin{aligned}\frac{dF}{d\theta} &= -\phi \\ \frac{dF}{dv} &= -p\end{aligned}\right\} \quad \ldots\ldots\ldots\ldots\ldots\ldots (120).$$

By means of these results we can easily express every quantity referring to the substance in terms of (θ, v, F) and the differential coefficients of F with respect to θ and v. Thus we have

$$U = F + \theta\phi$$
$$= F - \theta\frac{dF}{d\theta} \quad \ldots\ldots\ldots\ldots\ldots\ldots (121).$$

Also $\quad C_v = \dfrac{d_v U}{d\theta} = -\theta\dfrac{d^2 F}{d\theta^2} \quad \ldots\ldots\ldots\ldots (122),$

and since $\quad dQ = C_v d\theta + \theta\dfrac{d_v p}{d\theta} dv,$

THE THERMODYNAMIC POTENTIAL.

we obtain
$$C_p = C_v + \theta \frac{d_v p}{d\theta} \frac{d_p v}{d\theta}$$

$$= C_v - \theta \frac{\left(\frac{d_v p}{d\theta}\right)^2}{\frac{d_\theta p}{dv}}$$

$$= -\theta \frac{d^2 \mathcal{F}}{d\theta^2} + \theta \frac{\left(\frac{d^2 \mathcal{F}}{dv d\theta}\right)^2}{\frac{d^2 \mathcal{F}}{dv^2}} \quad \ldots\ldots\ldots\ldots (123).$$

Again, if K_θ be the isothermal compressibility and e the coefficient of cubical dilatation by heat at constant pressure,

$$K_\theta = -\frac{1}{v}\frac{d_\theta v}{dp} = -\frac{1}{v\,\frac{d_\theta p}{dv}}$$

$$= \frac{1}{v\,\frac{d^2 \mathcal{F}}{dv^2}} \quad \ldots\ldots\ldots\ldots (124),$$

$$e = \frac{1}{v}\frac{d_p v}{d\theta} = -\frac{\frac{d_v p}{d\theta}}{v\,\frac{d_\theta p}{dv}}$$

$$= -\frac{\frac{d^2 \mathcal{F}}{d\theta dv}}{v\,\frac{d^2 \mathcal{F}}{dv^2}} \quad \ldots\ldots\ldots\ldots (125).$$

When θ and p can be taken as independent variables, we have
$$d\Phi = dU - \theta d\phi - \phi d\theta + p dv + v dp,$$

or, since $dU = dQ - pdv$ and in a state of stable equilibrium $dQ = \theta d\phi$,

$$d\Phi = -\phi d\theta + vdp,$$

and therefore
$$\left.\begin{array}{l}\dfrac{d\Phi}{d\theta} = -\phi \\[1ex] \dfrac{d\Phi}{dp} = v\end{array}\right\} \quad \ldots\ldots\ldots\ldots\ldots (126).$$

Hence also
$$U = \Phi - \theta\frac{d\Phi}{d\theta} - p\frac{d\Phi}{dp} \ldots\ldots\ldots\ldots(127),$$

$$C_p = \frac{d_p U}{d\theta} + p\frac{d_p v}{d\theta}$$

$$= -\theta\frac{d^2\Phi}{d\theta^2} \ldots\ldots\ldots\ldots\ldots\ldots(128),$$

and
$$C_v = C_p - \theta\frac{d_p v}{d\theta}\frac{d_v p}{d\theta}$$

$$= C_p + \theta\frac{\left(\dfrac{d_p v}{d\theta}\right)^2}{\dfrac{d_\theta v}{dp}}$$

$$= -\theta\frac{d^2\Phi}{d\theta^2} + \theta\frac{\left(\dfrac{d^2\Phi}{dpd\theta}\right)^2}{\dfrac{d^2\Phi}{dp^2}} \ldots\ldots\ldots(129).$$

Again
$$K_\theta = -\frac{1}{v}\frac{d_\theta v}{dp}$$

$$= -\frac{\dfrac{d^2\Phi}{dp^2}}{\dfrac{d\Phi}{dp}} \ldots\ldots\ldots\ldots\ldots(130),$$

THE THERMODYNAMIC POTENTIAL.

and
$$e = \frac{1}{v}\frac{d_p v}{d\theta}$$

$$= \frac{\dfrac{d^2\Phi}{dp\,d\theta}}{\dfrac{d\Phi}{dp}} \quad \ldots\ldots\ldots\ldots(131).$$

128. As an illustration, we will find $\mathcal{F}$ and Φ for unit mass of ideal perfect gas satisfying the relation $pv = R\theta$.

Since we have
$$\left.\begin{array}{l} dU = C_v d\theta \\ dU = dQ - p\,dv \end{array}\right\},$$

we find
$$d\phi = C_v \frac{d\theta}{\theta} + \frac{p\,dv}{\theta},$$

or
$$d\phi = C_v \frac{d\theta}{\theta} + R\frac{dv}{v}.$$

Hence
$$U - U_0 = C_v(\theta - \theta_0),$$

and
$$\phi - \phi_0 = C_v \log \frac{\theta}{\theta_0} + R \log \frac{v}{v_0}$$

$$= (C_v + R) \log \frac{\theta}{\theta_0} - R \log \frac{p}{p_0},$$

since
$$\frac{v}{v_0} = \frac{\theta}{\theta_0}\frac{p_0}{p}.$$

Thus
$$\left.\begin{array}{l} \mathcal{F} = U - \theta\phi \\ \quad = U_0 - C_v\theta_0 - \theta\phi_0 + C_v\theta\left(1 - \log\dfrac{\theta}{\theta_0}\right) - R\theta\log\dfrac{v}{v_0} \\ \Phi = U - \theta\phi + pv \\ \quad = U_0 - C_v\theta_0 - \theta\phi_0 + (C_v+R)\theta\left(1-\log\dfrac{\theta}{\theta_0}\right) + R\theta\log\dfrac{p}{p_0} \end{array}\right\}(132).$$

From these equations, taking θ and v as independent variables, we get

$$\frac{d\mathcal{F}}{d\theta} = -\left(\phi_0 + C_v \log\frac{\theta}{\theta_0} + R\log\frac{v}{v_0}\right) = -\phi,$$

$$\frac{d\mathcal{F}}{dv} = -\frac{R\theta}{v} = -p.$$

Taking θ and p as independent variables,

$$\frac{d\Phi}{d\theta} = -\left\{\phi_0 + (C_v + R)\log\frac{\theta}{\theta_0} - R\log\frac{p}{p_0}\right\} = -\phi,$$

$$\frac{d\Phi}{dp} = \frac{R\theta}{p} = v.$$

Now the equation $pv = R\theta$ shows that at a given temperature and pressure, the value of R for different perfect gases is proportional to v. Hence if D be the density of the gas, that is, the mass contained in a unit of volume, at a given temperature and pressure, say at $0°$ C. and at a pressure of one atmo, we have

$$R = \frac{k}{D},$$

where k is a constant which has the same value for all perfect gases. Equations (132) therefore become

$$\left.\begin{array}{l}\mathcal{F} = U_0 - C_v\theta_0 - \theta\phi_0 + C_v\theta\left(1 - \log\frac{\theta}{\theta_0}\right) - \frac{k\theta}{D}\log\frac{v}{v_0} \\ \Phi = U_0 - C_v\theta_0 - \theta\phi_0 + \left(C_v + \frac{k}{D}\right)\theta\left(1 - \log\frac{\theta}{\theta_0}\right) + \frac{k\theta}{D}\log\frac{p}{p_0}\end{array}\right\}$$

$$\dots\dots\dots\dots (133).$$

129. The methods of entropy and of the thermodynamic potential are combined in a beautiful geometrical construction due to Prof. W. Gibbs, which does not seem

THE THERMODYNAMIC POTENTIAL. 337

to have obtained the attention it appears to deserve. This, no doubt, is owing to the fact that W. Gibbs' works are practically inaccessible to European readers, the only abstract with which we are acquainted—in addition, of course, to the very brief notice in Maxwell's 'Theory of Heat'—being given in Duhem's treatise on the Thermodynamic Potential.

In Gibbs' geometrical method, which supposes the substance to be acted on by no external forces except a pressure on the surface which becomes uniform in a state

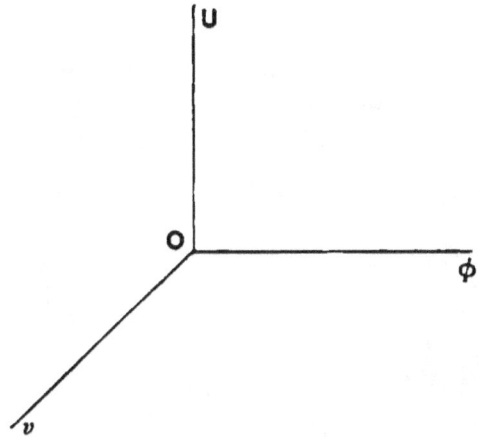

of equilibrium, we take three rectangular axes: the axis of x representing the entropy, the axis of y the volume and the axis of z the energy. The different states of equilibrium of the substance will then be represented by a surface.

Now at any point (x, y, z) of the surface, the direction-cosines of the normal are known to be proportional to

$$\left(\frac{d_y z}{dx}, \frac{d_x z}{dy}, -1\right),$$

that is, to
$$\left(\frac{d_v U}{d\phi}, \frac{d_\phi U}{dv}, -1\right).$$

But since any line drawn on the surface represents a reversible, or, at least, an equilibrium path, we have

$$\left.\begin{array}{l} d_v U = d_v Q = \theta d_v \phi \\ d_\phi U = - p dv \end{array}\right\}.$$

Thus $\dfrac{d_v U}{d\phi} = \theta$ and $\dfrac{d_\phi U}{dv} = -p.$

The direction-cosines of the normal are therefore proportional to $(\theta, -p, -1)$. Hence any two points of the surface at which the tangent planes are parallel correspond to two states in which the substance is in equilibrium at the same temperature and pressure.

We are now prepared to examine whether a state of equilibrium is stable. For let the substance to be considered be contained in a cylinder fitted with a smooth air-tight piston and placed within a closed vessel of

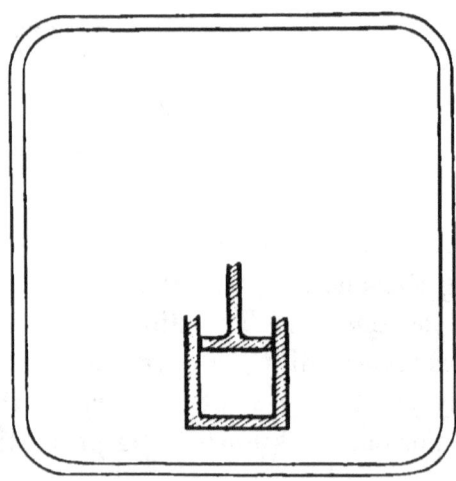

immense size, of invariable form and volume, covered with the best kind of non-conducting material. Then let the large vessel be filled with some fluid or gas whose state of equilibrium is always stable. If the fluid be light and compressible, like air, we shall have the case of the thermodynamic potential at constant pressure: if the fluid be incompressible, we obtain the case of the thermodynamic potential at constant volume. In what follows, we suppose the fluid to be air.

If, under these conditions, the substance can exist in any two different states of equilibrium whatever, it is obvious that the volume and the energy of the whole system will have the same value in both cases: also, on account of the immense size of the larger vessel, that the pressure and the temperature will have the same uniform values in both states.

The equilibrium in a state A will be stable, under the given conditions of pressure, if the system is unable, of itself, to pass into any other state P. Now the temperature in the state P will not generally be the same, at first, as in the state A; but by allowing the system to stand long enough and, if necessary, removing part of the non-conducting covering of the cylinder, it will always be possible, without violating the conditions of the system, to bring the system to a third state B in which the temperature has its original uniform value. We have, therefore, only to determine whether the system is able, of itself, to pass from the state A to any other state B at the same temperature and pressure as A.

Since the system is unable to receive or lose heat on account of the non-conducting covering, the substance will be unable, of itself, to pass from A to B if the entropy of the whole system be greater in the state A than in the state B. To exhibit this condition geometrically, let the entropy, volume and energy of the fluid which surrounds the cylinder be represented by three rectangular axes drawn in opposite directions to the former set of axes. Then since the quantity of fluid is very great, all its states of equilibrium will be represented by a plane surface. Also, if we choose the origins of coordinates such that when the substance within the cylinder is in equilibrium with the surrounding fluid at any given pressure and temperature, the corresponding states are represented by two points lying on a line parallel to $O\phi$, it is clear that any other state of equilibrium will be represented by two points also lying on a line parallel to $O\phi$.

The accompanying figure is supposed to be a view of the surface as seen by an observer at a great distance on

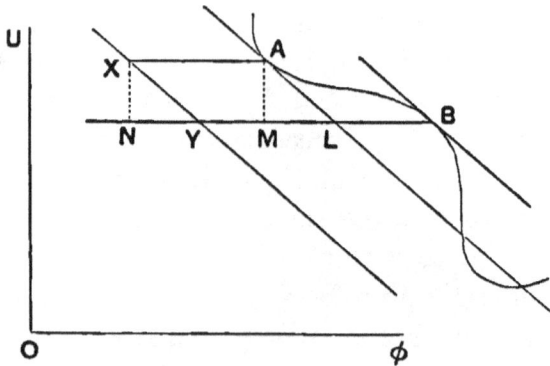

the axis Ov. The plane which represents the states of the external fluid is intersected in the points X, Y, by lines through A and B parallel to $O\phi$; and AM, XN are planes parallel to the plane UOv.

The increase of the entropy of the substance within the cylinder will therefore be proportional to BM and the decrease of the entropy of the rest of the system to NY.

The total increase of entropy is therefore proportional to
$MB - NY \equiv MB + MY - NM \equiv BY - AX \equiv BL$.

If, therefore, the point L be to the right of the point B, the substance will be unable to pass from the state A to the state B and the point A represents a stable con-

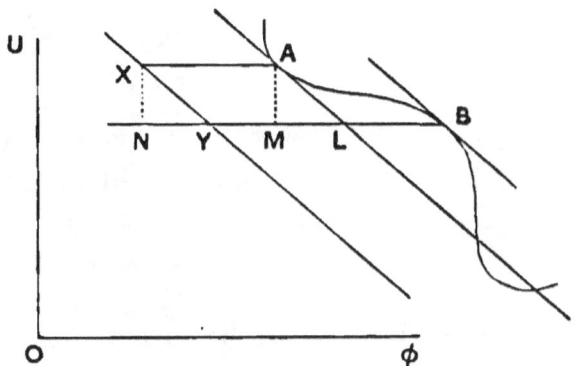

dition of the substance within the cylinder. On the other hand, if L be to the left of B, the point A will represent a state of the substance which, if possible for an instant, is essentially unstable and cannot be permanent.

Instances of unstable states occur 'when a liquid not in presence of its vapour is heated above its boiling point, and also when a liquid is cooled below its freezing point, or when a solution of a salt or a gas becomes supersaturated.

In the first of these cases, the contact of the smallest quantity of vapour will produce an explosive evaporation; in the second, the contact of ice will produce explosive freezing; in the third, a crystal of the salt will produce explosive crystallization; and in the fourth, a bubble of any gas will produce explosive effervescence.'

Again, when the surface touches a tangent plane in

two points A, B, and is to the left of it everywhere else, portions of the substance can permanently coexist at the

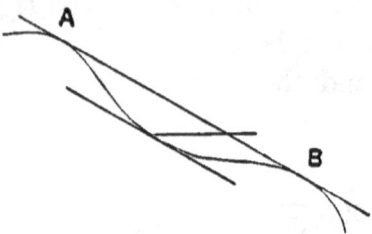

same temperature and pressure in the two states corresponding to the points A, B, and it will be possible to pass from one state to the other by a reversible operation during which the temperature and pressure remain constant.

If a mass x in the state A be in equilibrium with a mass $1-x$ in the state B, the corresponding point in the diagram will be a point in the line AB, coinciding with the centre of gravity of a mass x placed at A and a mass $1-x$ placed at B. Hence, since x may have all possible values between 0 and 1, every point in the limited line AB represents a condition of the substance when it is partly in the state A and partly in the state B. If now the tangent plane roll on the 'primitive' surface representing the state of the substance when homogeneous, the locus of the line AB will be another surface, called the 'secondary' surface, which represents the condition of the substance when part is in one state and part in another.

For the sake of fixing the ideas, let the points A, B, belong to the liquid and gaseous states, respectively. Then there will be two other rolling planes corresponding to the coexistence of the solid with the liquid, and of the

THE THERMODYNAMIC POTENTIAL. 343

solid with the gaseous, states. If the first of these touch the surface in the same point A_0 as the rolling plane AB, the two planes will then coincide, since the surface is known to be continuous and to possess only one tangent plane, at A_0. The tangent plane which touches the surface in the point A_0 will therefore touch in two other points B_0, C_0. In this case, the three rolling planes coincide: the physical interpretation being that the substance can then exist simultaneously in the three forms of solid, liquid and gaseous.

The three developable surfaces obtained by means of the three rolling planes, together with the plane triangle $A_0 B_0 C_0$, which corresponds to the 'triple point,' constitute what Prof. Gibbs calls the 'Surface of Dissipated Energy.'

CHAPTER VI.

APPLICATIONS OF THE THERMODYNAMIC POTENTIAL.

A. *Change of state of aggregation.*

130. THE questions which we now proceed to discuss were first examined by elementary methods. Some of the results have been already given. The rest are chiefly due to M. Moutier. The thermodynamic potential was first applied to the questions by Duhem in 1886, and they will be found to afford very simple illustrations of its use.

We suppose that two portions of the substance can coexist in stable equilibrium in two different states at the same temperature and pressure, and that, consequently, it is possible to pass from one state to the other by a reversible operation during which the temperature and pressure remain constant. There will also be states of unstable equilibrium which may be obtained thus. Beginning with a state of stable equilibrium and keeping the pressure constant, we endeavour to raise or lower the temperature; or, keeping the temperature constant, we endeavour to increase or diminish the pressure, without

APPLICATIONS OF THE THERMODYNAMIC POTENTIAL. 345

causing the substance to pass from one state to the other. We then wish to be able to distinguish between stable and unstable states of equilibrium when the system is contained in a vessel of any kind and subjected to no external forces but a uniform and constant normal surface pressure. For this purpose it seems natural to employ the thermodynamic potential at constant pressure. The fact that we are unable to take θ and p as independent variables to define the state of the system makes no difference to the two principal properties of the thermodynamic potential—that its variation is always negative and that, when it becomes a small quantity of the second order, the equilibrium is stable.

Let the system to be considered consist of a mass w in the state (w) and a mass s in the state (s), at the same temperature θ and pressure p. Also let the thermodynamic potential at constant pressure of unit mass of the substance in the first state be Φ_w and in the second state Φ_s: Φ_w and Φ_s both being functions of p and θ. Then if Φ be the thermodynamic potential at constant pressure of the whole system, we shall have

$$\Phi = w\Phi_w + s\Phi_s.$$

If a mass ds pass from the state (w) to the state (s) by any path whatever in which the external pressure is constantly equal to p and if the temperature be the same at the end of the process as at the beginning, we have, since Φ_w and Φ_s are unaltered,

$$d\Phi = (\Phi_s - \Phi_w)\, ds.$$

Now $d\Phi$ must be negative. Hence if Φ_s be greater than Φ_w, the system will be unable to pass from the state (w) to the state (s), but it may pass from (s) to (w). If Φ_s be less than Φ_w, the system will only be able to pass from the state (w) to the state (s). If Φ_s be equal to Φ_w, the equilibrium will be stable, the system being unable to pass either from (w) to (s) or from (s) to (w).

From the equation $\Phi_s = \Phi_w$, we see that p will be a function of θ when the two states of the substance are in stable equilibrium together at the same temperature and pressure. This conclusion has already been drawn in the case of the solid and liquid states, and might have been proved in the same way in other cases.

If $(p+dp,\ \theta+d\theta)$ be consecutive values of $(p,\ \theta)$ satisfying the equation $\Phi_s = \Phi_w$, we obtain

$$\frac{d\Phi_s}{dp}dp + \frac{d\Phi_s}{d\theta}d\theta = \frac{d\Phi_w}{dp}dp + \frac{d\Phi_w}{d\theta}d\theta.$$

But we may take θ and p as independent variables when the substance is entirely in one state. Hence if the volume and entropy of unit mass in the two states be denoted by the symbols $(v_w,\ \phi_w)$ and $(v_s,\ \phi_s)$, respectively, we get

$$v_s dp - \phi_s d\theta = v_w dp - \phi_w d\theta,$$

and therefore
$$\phi_s - \phi_w = (v_s - v_w)\frac{dp}{d\theta},$$

$\dfrac{dp}{d\theta}$ being found from the equation $\Phi_s = \Phi_w$.

If L be the latent heat of transition from the state (w) to the state (s), we have

$$\frac{L}{\theta} = \phi_s - \phi_w,$$

and thence

$$L = \theta(v_s - v_w)\frac{dp}{d\theta} \quad\ldots\ldots\ldots\ldots\therefore (134).$$

APPLICATIONS OF THE THERMODYNAMIC POTENTIAL. 347

Also if we take two rectangular axes, and measure temperature along Ox and pressure along Oy, the equation $\Phi_s = \Phi_w$ will represent a curve on opposite sides of which $\Phi_s - \Phi_w$ will have different signs.

For if a straight line be drawn parallel to $O\theta$, cutting the curve in a point P whose coordinates are (θ, p), the

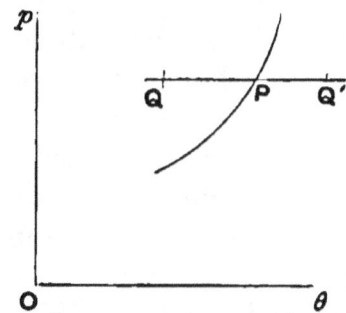

coordinates of a point Q a little to the left may be written $(\theta - d\theta, p)$, and

$$\Phi_s(\theta - d\theta, p) - \Phi_w(\theta - d\theta, p) = \Phi_s - d\theta \frac{d\Phi_s}{d\theta} - \Phi_w + d\theta \frac{d\Phi_w}{d\theta}$$

$$= -d\theta \left(\frac{d\Phi_s}{d\theta} - \frac{d\Phi_w}{d\theta} \right)$$

$$= d\theta (\phi_s - \phi_w)$$

$$= \frac{d\theta L}{\theta}.$$

The coordinates of a point a little to the right will be $(\theta + d\theta, p)$, and

$$\Phi_s(\theta + d\theta, p) - \Phi_w(\theta + d\theta, p) = \Phi_s + d\theta \frac{d\Phi_s}{d\theta} - \Phi_w - d\theta \frac{d\Phi_w}{d\theta}$$

$$= d\theta \left(\frac{d\Phi_s}{d\theta} - \frac{d\Phi_w}{d\theta} \right)$$

$$= -\frac{d\theta L}{\theta}.$$

The equation $d\Phi = (\Phi_s - \Phi_w) ds$ then becomes

$$d\Phi = \frac{d\theta L ds}{\theta}$$

in the first case, and

$$d\Phi = -\frac{d\theta L ds}{\theta}$$

in the second.

Hence we conclude that the only event possible on the right of the curve is that which absorbs heat; and on the left of the curve, that by which heat is evolved. For example, if a vessel contain water and saturated steam, and we can manage to raise the temperature without altering the external pressure or causing the water to evaporate, the only phenomena possible will be evaporation of the water and consequent absorption of heat. But if the temperature be reduced, the only phenomena possible will be condensation of steam and evolution of heat.

Again, if a line be drawn through P parallel to OP, the coordinates of a point, R, a little above P may be written $(\theta, p + dp)$, and

$$\Phi_s(\theta, p+dp) - \Phi_w(\theta, p+dp) = \Phi_s + dp\frac{d\Phi_s}{dp} - \Phi_w - dp\frac{d\Phi_w}{dp}$$

$$= (v_s - v_w) dp.$$

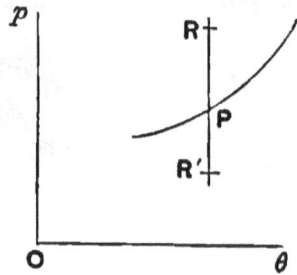

APPLICATIONS OF THE THERMODYNAMIC POTENTIAL. 349

The coordinates of a point, R', a little below P will be $(\theta, p - dp)$, and

$$\Phi_s(\theta, p - dp) - \Phi_w(\theta, p - dp) = \Phi_s - dp\frac{d\Phi_s}{dp} - \Phi_w + dp\frac{d\Phi_w}{dp}$$

$$= -(v_s - v_w)\,dp.$$

Hence at all points above the curve, the only event possible is a change to the state of smaller volume; below the curve, to the state of greater volume. Thus if a vessel contain water and saturated steam, and we succeed in increasing the pressure without altering the temperature or causing the steam to condense, the water will be unable to evaporate but the steam will be in danger of liquefying. If we reduce the pressure without altering the temperature or causing the water to evaporate, the steam will be unable to condense and the water will be in danger of explosive evaporation.

131. If the substance which can exist in two different states is contained in a vessel of given volume, we must use the thermodynamic potential at constant volume.

The thermodynamic potential at constant volume of unit mass in the state (w) will be a function of θ and v_w which may be denoted by $\mathcal{F}_w(\theta, v_w)$. In like manner, the thermodynamic potential at constant volume of unit mass in the state (s), at the same temperature θ, may be written $\mathcal{F}_s(\theta, v_s)$. Thus, if $\mathcal{F}$ be the thermodynamic potential at constant volume of the whole system,

$$\mathcal{F} = w\mathcal{F}_w(\theta, v_w) + s\mathcal{F}_s(\theta, v_s).$$

Hence, if a mass ds pass from the state (w) to the state (s) by any path whatever during which the volume is constant, and if the temperature be the same at the end of the process as at the beginning, we shall have

$$d\mathcal{F} = (\mathcal{F}_s - \mathcal{F}_w)\, ds + s\, \frac{d\mathcal{F}_s}{dv_s}\, dv_s + w\, \frac{d\mathcal{F}_w}{dv_w}\, dv_w.$$

Now in a state of stable equilibrium, we have $d\mathcal{F} = 0$: also $\dfrac{d\mathcal{F}_s}{dv_s} = \dfrac{d\mathcal{F}_w}{dv_w} = -p$. Thus

$$(\mathcal{F}_s - \mathcal{F}_w)\, ds - p\, (s\, dv_s + w\, dv_w) = 0.$$

But since $sv_s + wv_w$ and $s + w$ are both constant, we have

$$s\, dv_s + w\, dv_w = -(v_s - v_w)\, ds.$$

Hence in a state of stable equilibrium,

$$\mathcal{F}_s - \mathcal{F}_w + p\, (v_s - v_w) = 0,$$

or, since $\quad \Phi_s = \mathcal{F}_s + pv_s$ and $\Phi_w = \mathcal{F}_w + pv_w$,

$$\Phi_s = \Phi_w,$$

which is the very same equation as was found to hold when the external pressure was constant. The conditions and properties of stable equilibrium are therefore the same whether the pressure or the volume be invariable.

132. If the substance can exist in a third state (i), the condition that the state (s) may be in stable equilibrium with the state (i) is $\Phi_s = \Phi_i$. Hence when the same values (p', θ') of (p, θ) satisfy the two equations $\Phi_s = \Phi_w$ and $\Phi_s = \Phi_i$, we have

$$\left.\begin{array}{l}\Phi_s(p', \theta') = \Phi_w(p', \theta') \\ \Phi_s(p', \theta') = \Phi_i(p', \theta')\end{array}\right\}.$$

From these two equations, we obtain

$$\Phi_w(p', \theta') = \Phi_i(p', \theta').$$

APPLICATIONS OF THE THERMODYNAMIC POTENTIAL. 351

We therefore see that the two states (w) and (i) can also exist permanently together at the pressure p' and temperature θ', which thus constitutes a triple point.

If we take two rectangular axes to denote temperature and pressure, we shall have three curves corresponding to the three equations

$$\left.\begin{array}{l}\Phi_s = \Phi_w \\ \Phi_s = \Phi_i \\ \Phi_w = \Phi_i\end{array}\right\}.$$

These curves are such that if any point be common to two of them, the third will also pass through the same point. But we know that, in general, there is but one such point. The relative position of the three curves will therefore be known if we are able to determine their distribution in the neighbourhood of this, the triple, point.

At any given temperature θ, let us suppose that the equations to the three curves give the values (p_1, p_2, p_3), respectively, for p. Then when θ is very near to θ', we obtain

$$\Phi_s(p_1, \theta) = \Phi_w(p_1, \theta),$$

$$\Phi_s(p_1, \theta) + (p_2 - p_1)\frac{d\Phi_s(p_1, \theta)}{dp_1}$$
$$= \Phi_i(p_1, \theta) + (p_2 - p_1)\frac{d\Phi_i(p_1, \theta)}{dp_1},$$

$$\Phi_w(p_1, \theta) + (p_3 - p_1)\frac{d\Phi_w(p_1, \theta)}{dp_1}$$
$$= \Phi_i(p_1, \theta) + (p_3 - p_1)\frac{d\Phi_i(p_1, \theta)}{dp_1}.$$

Hence

$$(p_2 - p_1)\left\{\frac{d\Phi_s(p_1, \theta)}{dp_1} - \frac{d\Phi_i(p_1, \theta)}{dp_1}\right\}$$
$$= (p_3 - p_1)\left\{\frac{d\Phi_w(p_1, \theta)}{dp_1} - \frac{d\Phi_i(p_1, \theta)}{dp_1}\right\}.$$

If the volumes of unit mass in the three states be denoted at the triple point by (v_s', v_w', v_i'), respectively, we get, by rejecting small quantities of the second order,

$$(p_2 - p_1)(v_s' - v_i') = (p_3 - p_1)(v_w' - v_i').$$

Similarly,

$$(p_3 - p_2)(v_w' - v_i') = (p_1 - p_2)(v_w' - v_s'),$$
$$(p_1 - p_3)(v_s' - v_w') = (p_2 - p_3)(v_s' - v_i').$$

In the case of water in the three forms of steam, water and ice, the volumes arranged in order of magnitude are known to be (v_s, v_i, v_w).

Also at a temperature a little below θ', the greatest pres-

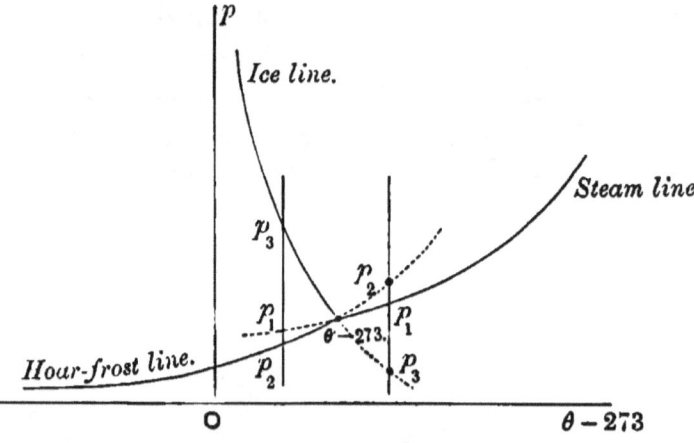

sure is known to be p_3. The three pressures arranged in order of magnitude are therefore (p_2, p_1, p_3).

More generally, if a line be drawn parallel to Op, intersecting the curves in three points, the middle point will belong to that curve which corresponds to the greatest change of volume.

B. *Saline Solutions.*

133. The vapour emitted by dilute sulphuric acid, or an aqueous solution of a salt, is known to be pure aqueous vapour. We also know that a solution can be in equilibrium with the salt but not with water so that there is a maximum but not a minimum limit to the strength of the solution. We now propose to examine some of the properties of solutions. The methods of the thermodynamic potential were first employed for this purpose by Helmholtz in some memoirs on electrical subjects which have been translated by the Physical Society. A comprehensive and simple investigation has also appeared in Duhem's work on the 'Thermodynamic Potential'.

The systems to be considered are supposed to be subjected to no external forces but a uniform and constant normal surface pressure, and we therefore make use of the thermodynamic potential at constant pressure.

Let a homogeneous solution in a state of equilibrium be composed of a mass s of salt and a mass w of water. Also let the constant surface pressure be p and the uniform temperature of the solution θ. Then Φ_{sw}, the thermodynamic potential at constant pressure of the solution will be a function of the four variables

$$(\theta, p, s, w).$$

Now if λ be any positive quantity, and we take a second homogeneous solution at the same pressure and temperature and of the same composition as the first but λ times as great, its thermodynamic potential at constant pressure will evidently be $\lambda \Phi_{sw}$. Thus when we multiply both s and w by the same factor λ without altering θ or p, we also multiply Φ_{sw} by λ. We therefore conclude that as far as the variables s and w are concerned, Φ_{sw} is a homogeneous function of the first degree. Hence if we put

$$\frac{d\Phi_{sw}}{ds} = F_s, \quad \frac{d\Phi_{sw}}{dw} = F_w,$$

we obtain, by Euler's theorem of homogeneous functions,

$$\Phi_{sw} = sF_s + wF_w \quad \ldots\ldots\ldots\ldots(135).$$

The functions F_s and F_w being evidently homogeneous and of the degree 0 in s and w, we have again by Euler's theorem,

$$\left. \begin{array}{l} s\dfrac{dF_s}{ds} + w\dfrac{dF_s}{dw} = 0 \\[4pt] s\dfrac{dF_w}{ds} + w\dfrac{dF_w}{dw} = 0 \end{array} \right\},$$

or, since
$$\frac{dF_s}{dw} = \frac{dF_w}{ds} = \frac{d^2\Phi_{sw}}{ds\,dw},$$

$$\left. \begin{array}{l} s\dfrac{dF_s}{ds} + w\dfrac{dF_w}{ds} = 0 \\[4pt] s\dfrac{dF_s}{dw} + w\dfrac{dF_w}{dw} = 0 \end{array} \right\} \quad \ldots\ldots\ldots(136).$$

134. If a system be composed of the homogeneous saline solution just considered together with a mass μ_s of free salt at the bottom of the solution and a mass μ_w of water-vapour above it, in equilibrium at the pressure p and temperature θ, the thermodynamic potential at constant pressure of the system will be

$$\Phi = sF_s + wF_w + \mu_s\Psi_s + \mu_w\Psi_w,$$

where Ψ_s and Ψ_w are the thermodynamic potentials at

APPLICATIONS OF THE THERMODYNAMIC POTENTIAL. 355

constant pressure of unit masses of salt and water-vapour, respectively, at the pressure p and temperature θ.

If a small change of state occur in consequence of which s increases by ds, μ_s will at the same time decrease by ds. Thus if θ, p and w are not affected by the change of state, we have, by equation (110),

$$d\Phi = (F_s - \Psi_s)\,ds + \left(s\frac{dF_s}{ds} + w\frac{dF_w}{ds}\right)ds$$
$$= (F_s - \Psi_s)\,ds.$$

Hence, when the solution is in stable equilibrium with salt, in which case it is said to be 'saturated',

$$F_s = \Psi_s \quad \ldots\ldots\ldots\ldots\ldots\ldots(137).$$

If we write h for $\dfrac{s}{w}$, the 'strength' of the solution, this result becomes a relation between (θ, p, h) which shows that the strength of a saturated solution depends only on the temperature and pressure.

In like manner, when the solution is in equilibrium with aqueous vapour, we find

$$F_w = \Psi_w \ldots\ldots\ldots\ldots\ldots\ldots(138),$$

from which we conclude that the pressure of the aqueous vapour which is in stable equilibrium with a solution depends only on the temperature and on the strength of the solution.

135. Now suppose that, at a given pressure p and temperature θ, we have two solutions, the first of which is formed by dissolving a mass $s+ds$ of salt in a mass w of water, and the second by dissolving a mass $s-ds$ of salt in another mass w of water: also suppose the pressure and temperature to be such that the solutions can give up neither salt nor steam. Then, on mixing the two solutions together, we get a new solution composed of a mass $2s$ of salt dissolved in a mass $2w$ of water.

The thermodynamic potential at constant pressure of the two solutions, before being mixed, is

$$\Phi(\theta, p, s+ds, w) + \Phi(\theta, p, s-ds, w),$$

and after mixing, if the temperature and pressure be the same as before,

$$\Phi(\theta, p, 2s, w), \text{ or } 2\Phi(\theta, p, s, w).$$

The operation being irreversible, the thermodynamic potential will have decreased. Hence

$$2\Phi(\theta, p, s, w) < \Phi(\theta, p, s+ds, w) + \Phi(\theta, p, s-ds, w),$$

or
$$2\Phi < \left\{\Phi + ds\frac{d\Phi}{ds} + \frac{(ds)^2}{2}\frac{d^2\Phi}{ds^2} + \ldots\right\}$$
$$+ \left\{\Phi - ds\frac{d\Phi}{ds} + \frac{(ds)^2}{2}\frac{d^2\Phi}{ds^2} + \ldots\right\}$$
$$< 2\Phi + (ds)^2 \frac{d^2\Phi}{ds^2} + \ldots$$

If we remember that $\dfrac{d\Phi}{ds} = F_s$, this inequality shows that $\dfrac{dF_s}{ds}$ is positive. But since F_s is a function of (θ, p, h) where $h = \dfrac{s}{w}$, we have $\dfrac{dF_s}{ds} = \dfrac{1}{w}\dfrac{dF_s}{dh}$. Hence $\dfrac{dF_s}{dh}$ is positive and F_s continually increases with the strength of the solution.

In like manner we may prove that $\dfrac{dF_w}{dw}$ is positive; and since $\dfrac{dF_w}{dw} = -\dfrac{s}{w^2}\dfrac{dF_w}{dh}$, we conclude that F_w con-

APPLICATIONS OF THE THERMODYNAMIC POTENTIAL. 357

tinually decreases as the strength of the solution increases. This result may also be deduced from the former, for, by equation (136), we have

$$h\frac{dF_s}{ds} + \frac{dF_w}{ds} = 0,$$

and therefore $h\dfrac{dF_s}{dh} + \dfrac{dF_w}{dh} = 0$. Hence if $\dfrac{dF_s}{dh}$ be positive, $\dfrac{dF_w}{dh}$ must be negative.

If we take two rectangular axes to denote temperature and pressure, equation (138) will give a series of curves for different values of h representing the relation between the vapour-pressure and the temperature and no two of these curves will intersect. For, if possible, let the curves belonging to two different values of h intersect in the point (θ, p). Then, by equation (138) we have

$$\left.\begin{array}{l}F_w(\theta, p, h) = \Psi_w(\theta, p)\\ F_w(\theta, p, h') = \Psi_w(\theta, p)\end{array}\right\},$$

so that $\qquad F_w(\theta, p, h) = F_w(\theta, p, h').$

But F_w continually decreases as h increases and therefore cannot have the same value for two different values of h.

Hence if a curve be drawn to express the relation between the pressure and temperature of saturated steam, which will coincide with the curve $F_w(\theta, p, h) = \Psi_w(\theta, p)$ when h is zero, the whole series of curves will lie entirely on one side of it, further and further away as h increases from zero. To determine their relative positions it will therefore be sufficient to know the position of the curve for which h is very small with respect to the curve for which h is zero. To find this, we may proceed as follows.

Differentiating equation (138) on the supposition that θ is constant, we get

$$\frac{dF_w}{dh} + \frac{dF_w}{dp}\frac{d_\theta p}{dh} = \frac{d_\theta \Psi_w}{dp}\frac{d_\theta p}{dh}.$$

But if v_w be the volume of unit mass of aqueous vapour in stable equilibrium with the solution, we have

$$\frac{d_\theta \Psi_w}{dp} = v_w.$$

Hence
$$\frac{dF_w}{dh} = \left(v_w - \frac{dF_w}{dp}\right)\frac{d_\theta p}{dh} \quad \ldots\ldots\ldots\ldots(139).$$

Now in obtaining the formula which gives the volume in a state of stable equilibrium in terms of the thermodynamic potential at constant pressure in the form

$$vdp = \Phi(p+dp) - \Phi(p),$$

we made no supposition except that the temperature remains constant and that a small change of p occasions only a small change of state. Thus, if v_s be the volume of unit mass of the solution, we have

$$(s+w)v_s dp = d(sF_s + wF_w),$$

where, on the right hand side, we may either take h constant or variable. If we wish to take h variable, let a small *arbitrary* quantity of vapour, $d\mu_w$, be formed, in consequence of which w becomes $w - d\mu_w$. Then

$$d(sF_s + wF_w) = \frac{d}{dp}(sF_s + wF_w)\,dp$$
$$\quad - \left(s\frac{dF_s}{dw} + w\frac{dF_w}{dw}\right)d\mu_w - (F_w - \Psi_w)\,d\mu_w$$
$$= \frac{d}{dp}(sF_s + wF_w)\,dp,$$

APPLICATIONS OF THE THERMODYNAMIC POTENTIAL. 359

by equations (136) and (138). Therefore

$$(s+w)v_s = s\frac{dF_s}{dp} + w\frac{dF_w}{dp},$$

or

$$(h+1)v_s = h\frac{dF_s}{dp} + \frac{dF_w}{dp}.$$

Thus as h tends to zero, $\frac{dF_w}{dp}$ will approximate to the volume of unit mass of water. This being less than v_w, we see from equation (139), that when h is small, $\frac{dF_w}{dh}$ and $\frac{d_\theta p}{dh}$ are of the same sign, so that $\frac{d_\theta p}{dh}$ is negative. Hence the curves all lie *below* the curve for which $h = 0$,

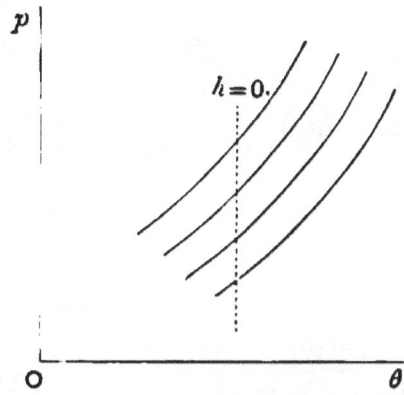

and therefore at any given temperature the vapour-pressure continually decreases as the strength of the solution increases.

136. If a solution contained in a vessel in a state of stable equilibrium be subjected to such a pressure that the vapour is just on the point of forming, and if, by slowly diminishing the pressure, a small amount of vapour $d\mu_w$, be formed in a reversible manner, a small quantity of heat which may be written $Ld\mu_w$, will be required to keep the temperature constant during the process. This quantity satisfies the equation

$$Ld\mu_w = \theta(\phi_2 - \phi_1),$$

where the suffixes refer to the two states of the system. But when the pressure is kept constant and a small change of temperature can only produce a small change of state, we have, when the pressure remains constant,

$$-\phi d\theta = d\Phi.$$

If, for convenience, we suppose θ to be the only quantity that varies, this result takes the simple form $-\phi = \dfrac{d\Phi}{d\theta}$.

Hence
$$\frac{Ld\mu_w}{\theta} = s\frac{dF_s(\theta, p, s, w)}{d\theta} + w\frac{dF_w(\theta, p, s, w)}{d\theta}$$
$$- s\frac{dF_s(\theta, p, s, w - d\mu_w)}{d\theta} - (w - d\mu_w)\frac{dF_w(\theta, p, s, w - d\mu_w)}{d\theta}$$
$$- d\mu_w \frac{d\Psi_w(\theta, p)}{d\theta},$$

and therefore
$$\frac{L}{\theta} = \frac{d}{d\theta}\left\{ s\frac{dF_s(\theta, p, s, w)}{dw} + w\frac{dF_w(\theta, p, s, w)}{dw} + F_w(\theta, p, s, w) \right.$$
$$\left. - \Psi_w(\theta, p) \right\}$$
$$= \frac{d}{d\theta}(F_w - \Psi_w), \text{ by equation (136)}.$$

Also by differentiating the equation $F_w = \Psi_w$, we get
$$\frac{d}{d\theta}(F_w - \Psi_w) = -\frac{d(F_w - \Psi_w)}{dp}\frac{d_h p}{d\theta}.$$

APPLICATIONS OF THE THERMODYNAMIC POTENTIAL. 361

Thus finally
$$L = \theta \left(v_w - \frac{dF_w}{dp}\right) \frac{d_h p}{d\theta} \quad \ldots\ldots\ldots\ldots(140).$$

Since $\frac{dF_w}{dp}$ will be small in comparison with v_w, we obtain approximately

$$L = \theta v_w \frac{d_h p}{d\theta}.$$

137. Kirchhoff's formula may be easily deduced by means of the thermodynamic potential. For when the pressure is constant, the equation

$$dU = dQ - pdv$$

becomes
$$dQ = d(U + pv) \quad \ldots\ldots\ldots\ldots(141)$$

Now we have
$$\Phi = U - \theta\phi + pv,$$

and in a state of stable equilibrium

$$\frac{d\Phi}{d\theta} = -\phi.$$

Hence
$$\Phi - \theta \frac{d\Phi}{d\theta} = U + pv,$$

and equation (141) becomes

$$dQ = d\left(\Phi - \theta \frac{d\Phi}{d\theta}\right) \quad \ldots\ldots\ldots\ldots(142).$$

If the operation consist of adding a quantity $d\mu_w$ of water to a solution formed of a mass s of salt dissolved in a mass w of water, and if $\chi(\theta, p)$ be the thermodynamic potential at constant pressure of unit mass of water,

$$\Phi = sF_s + wF_w + d\mu_w\chi,$$

$$d\Phi = \left(s\frac{dF_s}{dw} + w\frac{dF_w}{dw}\right)d\mu_w + (F_w - \chi)d\mu_w = (F_w - \chi)d\mu_w.$$

Therefore
$$\frac{dQ}{d\mu_w} = F_w - \chi - \theta \frac{d(F_w - \chi)}{d\theta},$$

and thence
$$\frac{d}{dh}\left(\frac{dQ}{d\mu_w}\right) = \frac{dF_w}{dh} - \theta \frac{d^2 F_w}{dh\, d\theta}.$$

Now since a liquid is very little affected by pressure, the value of $\dfrac{dF_w}{dh}$ will be practically independent of p. But when the pressure is equal to that of the vapour of the solution, we have approximately, by equation (139),
$$\frac{dF_w}{dh} = v_w \frac{d_\theta p}{dh},$$
and the vapour is generally so near the state of the ideal perfect gas that we shall introduce no serious error by putting $pv_w = R\theta$, where R is a constant. Hence
$$\frac{dF_w}{dh} = \frac{R\theta}{p} \frac{d_\theta p}{dh}$$
$$= R\theta \frac{d_\theta \log p}{dh},$$
and consequently
$$\frac{d}{dh}\left(\frac{dQ}{d\mu_w}\right) = -R\theta^2 \frac{d^2 \log p}{dh\, d\theta}.$$

If the pressure of saturated steam at the temperature θ be denoted by P, Kirchhoff's formula for the heat absorbed is obtained on integration: thus
$$\frac{dQ}{d\mu_w} = R\theta^2 \frac{d}{d\theta}\left(\log \frac{P}{p}\right) \quad \ldots\ldots\ldots\ldots (143).$$

138. The principles of thermodynamics were first applied to the freezing of saline solutions in 1886 by Duhem in his work on the Thermodynamic Potential. His investigations may be easily completed by aid of Dr Guthrie's beautiful experimental researches which may be seen in the Philosophical Magazine, 1875—1876.

APPLICATIONS OF THE THERMODYNAMIC POTENTIAL. 363

When an aqueous solution of a salt is slowly reduced in temperature, it is supposed that the ice which freezes out is perfectly pure. This appears to be strictly established by experiment when the solution contains no floating impurities and the cooling is so slow that the liquid solution does not become entangled among the ice.

Let a system in a state of equilibrium be composed of a homogeneous solution formed of a mass s of salt dissolved in a mass w of water, together with a mass μ_i of pure ice and a mass μ_s of free salt, not in contact with each other or with the solution. Also let the thermodynamic potential at constant pressure of unit mass of pure ice be $\Psi_i(\theta, p)$. Then the thermodynamic potential at constant pressure of the whole system, at the same temperature θ and pressure p, is

$$\Phi = sF_s + wF_w + \mu_s\Psi_s + \mu_i\Psi_i.$$

If a small quantity of ice fall into the solution in consequence of which w increases to $w + dw$, μ_i will at the same time decrease by dw. The temperature being brought to its original value and s remaining constant, it is easily shown that

$$d\Phi = (F_w - \Psi_i)\, dw.$$

This is always negative. Hence the only phenomenon possible will be the formation of ice when $F_w - \Psi_i$ is positive, and the disappearance of ice when $F_w - \Psi_i$ is negative. The condition that the system may be in stable equilibrium when the ice is in contact with the solution is

$$F_w(\theta, p, h) - \Psi_i(\theta, p) = 0 \ \ldots\ldots\ldots (144),$$

an equation which admits but one real value of h when θ and p are given, since when θ and p are constant, F_w continually decreases as h increases.

In like manner, if s increase to $s + ds$, and (θ, p, w) be unchanged, we have

$$d\Phi = (F_s - \Psi_s)\,ds,$$

and therefore the condition that the solution may be in stable equilibrium when the solution and free salt are in contact, is

$$F_s - \Psi_s = 0,$$

which is easily seen to have but one real root in h.

Again, it is well known that at ordinary temperatures, when salt and ice or salt and snow are mixed together, a violent chemical action ensues by which the temperature is greatly reduced and a liquid solution formed. The thermodynamic potential at constant pressure of a solution will therefore be less than the sum of the thermodynamic potentials of the salt and ice from which it is formed. Thus

$$sF_s + wF_w < s\Psi_s + w\Psi_i.$$

Hence, when $F_s - \Psi_s = 0$, $F_w - \Psi_i$ is negative. As h decreases from this value, $F_s - \Psi_s$ becomes negative and $F_w - \Psi_i$ continually increases. It therefore follows that the value of h which makes $F_w - \Psi_i = 0$ is less than that which makes $F_s - \Psi_s = 0$. Also for any value of h between these two limits, $F_s - \Psi_s$ and $F_w - \Psi_i$ are both negative. The solution is then capable of dissolving either salt or ice, and is stable if neither salt nor ice be present.

139. If we take two rectangular axes of θ and p, equation (144) will give a series of curves for different values of h which show how the pressure must vary with the temperature when the solution is in stable equilibrium with pure ice to keep the concentration of the solution

APPLICATIONS OF THE THERMODYNAMIC POTENTIAL. 365

constant. No two of these curves can intersect, for if the curves belonging to two different values of h could pass through the same point (θ, p), we should have

$$F_w(\theta, p, h) = \Psi_i(\theta, p),$$
$$F_w(\theta, p, h') = \Psi_i(\theta, p),$$

which is impossible, since F_w continually decreases as h increases.

The curves will therefore all lie on one side of that for which h is zero. To find their relative dispositions, we differentiate equation (144) on the supposition that θ is constant: thus

$$\frac{dF_w}{dh} + \left(\frac{dF_w}{dp} - \frac{d\Psi_i}{dp}\right)\frac{d_\theta p}{dh} = 0.$$

But if v_i be the volume of unit mass of ice, we have

$$\frac{d\Psi_i}{dp} = v_i.$$

Hence
$$\frac{dF_w}{dh} = \left(v_i - \frac{dF_w}{dp}\right)\frac{d_\theta p}{dh}.$$

Also when we restrict ourselves to a curve for which h is indefinitely small, the equation

$$(h+1)v_s = h\frac{dF_s}{dp} + \frac{dF_w}{dp}$$

shows that $\dfrac{dF_w}{dp}$ is equal to the volume of unit mass of the solution, which is the same as v_w, the volume of unit mass of water. Thus

$$\frac{dF_w}{dh} = (v_i - v_w)\frac{d_\theta p}{dh}.$$

The factor $v_i - v_w$ being positive, because water expands in freezing, we see that when h is small, $\dfrac{d_\theta p}{dh}$ is negative.

All the curves therefore lie below the curve of fusion of pure ice, that is, below the 'ice line.' Hence when the

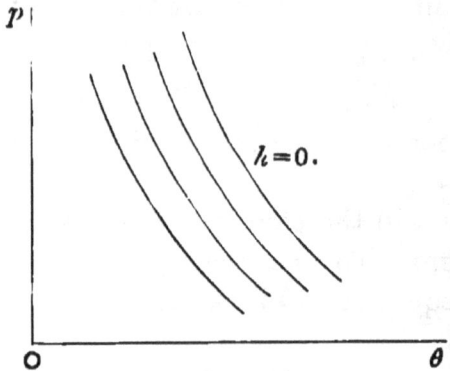

pressure is given, the temperature at which ice begins to separate out of a solution is lower for a strong than for a weaker solution.

When a salt is dissolved in a liquid, like acetic acid, which contracts in freezing, $v_i - v_w$ is negative, $\dfrac{dF_w}{dh}$ is also negative, as before, and therefore when $h = 0$, $\dfrac{d_\theta p}{dh}$ is positive. In this case, the curves corresponding to different values of h all lie above the curve for which $h = 0$. But on this

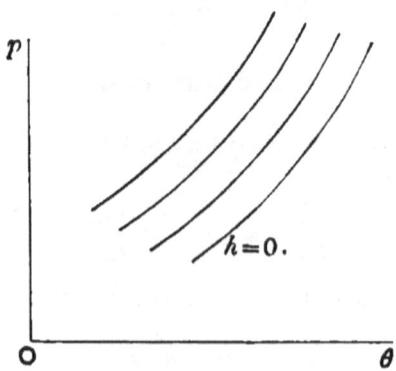

APPLICATIONS OF THE THERMODYNAMIC POTENTIAL. 367

curve, unlike the curve for water, θ and p increase together. Hence, as before, when the pressure is given, the temperature at which the pure frozen liquid begins to appear, is lower for a strong than for a weaker solution.

Thus in all cases, the presence of a salt retards the formation of ice from a liquid by cold, or $\dfrac{d_p h}{d\theta}$ is always negative for the stable solution of minimum strength.

140. If we suppose that every liquid can be reduced to the solid state by applying a sufficient degree of cold, it will follow that salt and ice are neutral to one another when the temperature is low enough. If a solution can then exist, it must be in stable equilibrium with both salt and ice simultaneously. The temperature at which this FIRST takes place and the strength of the corresponding solution, are given by the equations

$$\left. \begin{array}{l} F_w(\theta, p, h) = \Psi_i(\theta, p) \\ F_s(\theta, p, h) = \Psi_s(\theta, p) \end{array} \right\}.$$

If we take two rectangular axes to represent θ and h,

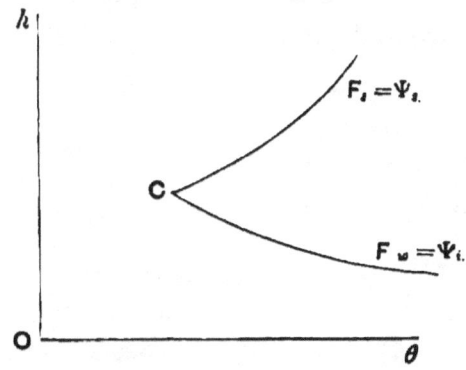

these equations will give two curves for each value of p, intersecting in a point C whose coordinates represent the temperature at which ice and salt first become neutral for that value of p, and the strength of the only stable solution which can then exist.

Let us take a mixture of ice and salt at the highest temperature at which they are neutral to one another, and then, by the application of heat, let a liquid solution be formed without altering either the temperature or the pressure. The operation, though an equilibrium operation, will be irreversible, for it is found that if we attempt to reduce the temperature, we do not dissociate the solution into its original salt and ice—what really occurs will be seen presently. If the solution be formed of a mass s of salt dissolved in a mass w of water, its thermodynamic potential at constant pressure, Φ, will be

$$\Phi = sF_s + wF_w,$$

or, since

$$\left.\begin{array}{l} F_s = \Psi_s \\ F_w = \Psi_i \end{array}\right\},$$

$$\Phi = s\Psi_s + w\Psi_i.$$

The thermodynamic potential is therefore unaltered by the process. But in any finite change of state in which the temperature and pressure remain constant, we have

$$\Delta\Phi = \Delta Q - \theta\Delta\phi.$$

The equation $\int_A^B \dfrac{dQ}{\theta} = \phi_B - \phi_A$ is thus *proved* to be true for this particular *irreversible* process.

141. It will now be necessary to introduce some additional experimental results, all of which are due to Dr Guthrie. They have only yet been shown to hold at atmospheric pressure, but we shall suppose them to be true at any pressure.

It is found that if we take a stable saline solution of any possible strength and subject it to reduction of temperature, the only solids which the solution gives up are pure ice or the anhydrous salt or a hydrate, until a certain definite temperature is reached which depends only on the nature of the salt and liquid which compose the solution, and not on their relative proportions. The solution then freezes, without further reduction of temperature, into a solid homogeneous brine. The temperature at which this takes place is found by Dr Guthrie to be the same as that at which ice and salt first become neutral to one another. Thence we infer that the final solution thus obtained, which Dr Guthrie calls a 'cryohydrate,' is of the same strength whatever may have been the strength of the solution with which we started, and that the lowest temperature attainable by means of a 'cryogen,' that is, a freezing mixture of salt and ice, is the freezing point of the corresponding cryohydrate.

These experimental conclusions lead to a very interesting irreversible non-frictional cycle which may be studied by means of strictly elementary methods.

Let the system consist originally of pure ice and a salt, such as nitrate of silver or chlorate of potassium, whose solution does not deposit a hydrate; or of ice and the hydrate of a salt like sulphate of magnesium. Also let the temperature θ_0 and pressure p_0 be such that θ_0 is the freezing point of the cryohydrate at the pressure p_0, and p_0 the pressure of the saturated vapour of ice at the temperature θ_0. Then, the materials being mixed together, the system may be made to undergo the following cyclical process at the constant temperature θ_0 and the constant pressure p_0.

(1) By slowly imparting heat, let a liquid cryohydrate be formed. Call the initial state of system A and the state at the end of this operation B: also denote the heat absorbed in the operation by Q_1.

The path AB is evidently irreversible, for by slowly abstracting heat, we should not dissociate the cryohydrate—we should simply cause it to freeze.

(2) By slowly increasing the volume, let the cryohydrate be evaporated at constant pressure and temperature, giving the saturated vapour of ice and the original salt or hydrate.

(3) Let the vapour be separated from the salt or hydrate, and then reduce it to ice.

The operations (2) and (3) are clearly reversible: hence if Q_2 be the heat absorbed during these processes, we have

$$Q_2 = \theta_0 (\phi_A - \phi_B).$$

But since the pressure is constant throughout the cycle, the total quantity of work done on the system during the cycle, and therefore also the total quantity of heat absorbed, is zero. Thus

$$Q_1 + Q_2 = 0.$$

Hence, for the irreversible path AB, in which the temperature is constantly equal to θ_0,

$$Q_1 = \theta_0 (\phi_B - \phi_A).$$

142. The subject we are studying may be simplified by means of a diagram. Take three rectangular axes, and let the axis of x denote the temperature, the axis of z the pressure, and the axis of y the strength of the solution. Then the weakest and strongest stable solutions will be

APPLICATIONS OF THE THERMODYNAMIC POTENTIAL. 371

represented in the diagram by two surfaces, given by the equations

$$\left. \begin{array}{l} \mathcal{F}_w(\theta, p, h) = \Psi_i(\theta, p) \\ \mathcal{F}_s(\theta, p, h) = \Psi_s(\theta, p) \end{array} \right\},$$

which meet along the freezing line of the cryohydrate, and all other stable solutions will be represented by points lying between these two surfaces.

The solutions which are in stable equilibrium with

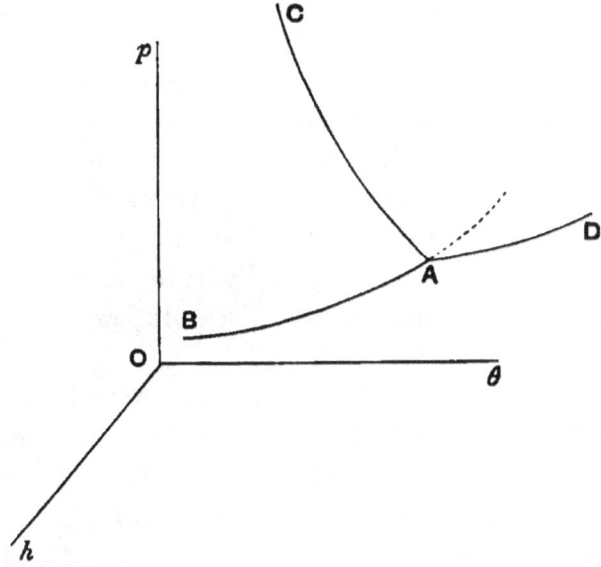

their own vapour will be represented by a surface close to the plane θOh, whose equation is

$$\mathcal{F}_w(\theta, p, h) = \Psi_w(\theta, p).$$

Every point below this surface represents a solution which is in danger of explosive evaporation.

The curve in which the surface $\mathcal{F}_w = \Psi_w$ meets the surface $\mathcal{F}_w = \Psi_i$, evidently refers to those solutions which are simultaneously in stable equilibrium with their own vapour and with pure ice. Its projection, AB, on the plane θOp, therefore denotes the relation between the temperature and the pressure of the saturated vapour of ice: in other words, it is the 'hoar-frost line' of water.

Again, if we draw another curve, AC, in the plane θOp, to represent the relation between the temperature and pressure when ice and water can exist in stable equilibrium together, the surface

$$\mathcal{F}_w(\theta, p, h) = \Psi_i(\theta, p).$$

will clearly meet the plane θOp in the curve AC and will there terminate, since h cannot be negative.

The freezing line of the cryohydrate may be called a 'quadruple line,' for four substances are in stable equilibrium with one another on it—pure ice, the anhydrous salt, the liquid cryohydrate and the solid cryohydrate. If the saturated solution yields a hydrate on being cooled, we must add a fifth to the list, viz., crystals of the hydrate.

The freezing line of the cryohydrate terminates in the point in which it meets the surface $\mathcal{F}_w = \Psi_w$. This point will always be a 'quintuple point,' and when the solution gives a hydrate, it will be a 'sextuple point.'

It will be noticed that since A is the 'triple point' of water, the curve AD in which $\mathcal{F}_w = \Psi_w$ meets the plane θOp, makes a finite angle with the projection on θOp of the curve of intersection of $\mathcal{F}_w = \Psi_w$ with $\mathcal{F}_w = \Psi_i$: but it should be remembered that there is no discontinuity or irregularity on the surface $\mathcal{F}_w = \Psi_w$ itself.

143. We have strong reasons for believing that h is constant along the freezing line of the cryohydrate. For, if possible, let the strength of the cryohydrate be greater

APPLICATIONS OF THE THERMODYNAMIC POTENTIAL. 373

at a point Q than at another point P of this line. Let the liquid cryohydrate belonging to P be frozen and then travel to the point Q by any path not lying between the surfaces

$$\left.\begin{array}{l}\mathcal{F}_s = \Psi_s \\ \mathcal{F}_w = \Psi_i \\ \mathcal{F}_w = \Psi_w\end{array}\right\}.$$

If the cryohydrate does not melt before the point Q is reached, it will then melt into two distinct parts, one of which is a liquid cryohydrate and the other pure solid ice. If it could melt before we get to Q, we should obtain a solution in the liquid form at a temperature below the freezing point of the cryohydrate at the same pressure. Both of these results are very improbable and, consequently, we conclude that h cannot have different values at two different points P, Q, of the freezing line of the cryohydrate.

A cryohydrate must therefore be regarded as a body of *definite chemical composition* (as was foreseen clearly by Dr Guthrie himself), and not as a mere solution in which the proportions of salt and ice depend on various accidents.

144. Any two consecutive points on the surface

$$\mathcal{F}_w(\theta, p, h) = \Psi_i(\theta, p)$$

satisfy the relation

$$dh = \frac{d_\theta h}{dp} dp + \frac{d_p h}{d\theta} d\theta,$$

where the differential coefficients $\frac{d_\theta h}{dp}$ and $\frac{d_p h}{d\theta}$ are to be found from the equation $\mathcal{F}_w(\theta, p, h) = \Psi_i(\theta, p)$.

If the two points also lie on the freezing line of the cryohydrate, we have $dh = 0$, and therefore, along this line,

$$\frac{d_\theta h}{dp} dp + \frac{d_p h}{d\theta} d\theta = 0.$$

Now we have already shown that when the salt is dissolved in a liquid which expands in the act of freezing, like water, the equation $\mathcal{F}_w(\theta, p, h) = \Psi_i(\theta, p)$ makes $\dfrac{d_\theta h}{dp}$ negative. If, however, the solution is formed by dissolving the salt in a liquid, like acetic acid, which contracts whilst freezing, then $\dfrac{d_\theta h}{dp}$ is positive. Also in both cases, $\dfrac{d_p h}{d\theta}$ is negative.

Hence along the freezing line of the cryohydrate, $\dfrac{dp}{d\theta}$ is negative in the former case and positive in the latter. Thus the freezing point of the cryohydrate is depressed or raised by pressure according as the liquid employed expands or contracts in the act of freezing. In other words, when the liquid employed in forming the solution is such that its freezing point is lowered by pressure, the freezing point of the cryohydrate is also lowered by pressure; and when the liquid employed is such that its freezing point is raised by pressure, the freezing point of the cryohydrate is also raised by pressure.

Again, let us suppose that at any point (θ, p) on the freezing line of the cryohydrate, the volume, energy, entropy and thermodynamic potential at constant pressure of unit mass of the cryohydrate are represented in the liquid state by the symbols (v', U', ϕ', Φ'), and in the solid state by $(v'', U'', \phi'', \Phi'')$, respectively: also let (v, U, ϕ, Φ)

APPLICATIONS OF THE THERMODYNAMIC POTENTIAL.

be respectively the sums of the volumes, energies, entropies and thermodynamic potentials at the same temperature and pressure of the salt and ice of which the cryohydrate is composed. Then by Art. 140 we have

$$\Phi = \Phi'.$$

Also since the freezing of a cryohydrate is a reversible process performed at constant temperature and pressure,

$$\Phi' = \Phi''.$$

Thus, at the point (θ, p),

$$\Phi = \Phi' = \Phi''.$$

At a consecutive point $(\theta + d\theta, p + dp)$, we have

$$d\Phi = d\Phi' = d\Phi'',$$

that is, $vdp - \phi d\theta = v'dp - \phi'd\theta = v''dp - \phi''d\theta.$

Hence
$$\left. \begin{array}{l} \phi' - \phi = (v' - v)\dfrac{dp}{d\theta} \\ \phi'' - \phi' = (v'' - v')\dfrac{dp}{d\theta} \end{array} \right.,$$

where $\dfrac{dp}{d\theta}$ refers to the freezing line of the cryohydrate.

But if L_1 be the heat absorbed in the formation of unit mass of the liquid cryohydrate from salt and ice at constant temperature and pressure, then $L_1 = \theta(\phi' - \phi)$: also if L_2 be the heat evolved in the freezing of the cryohydrate, $L_2 = \theta(\phi' - \phi'')$. Thus finally

$$\left. \begin{array}{l} L_1 = \theta(v' - v)\dfrac{dp}{d\theta} \\ L_2 = -\theta(v'' - v')\dfrac{dp}{d\theta} \end{array} \right\} \quad \ldots\ldots\ldots\ldots(145),$$

L_1 and L_2 both being positive.

An equation similar to the first of these will hold when the liquid solution is formed by the melting of ice with a hydrate.

Now when the liquid employed expands in freezing, like water, $\dfrac{dp}{d\theta}$ is negative along the freezing line of the cryohydrate: hence, by equations (145), the cryohydrate also expands in freezing, and when we form the liquid cryohydrate from solid ice and salt, or from solid ice and a hydrate, at constant temperature and pressure, there will be a contraction of volume. On the contrary, when the liquid employed is one which shrinks in freezing, the cryohydrate will also shrink in freezing, and when it is formed at constant temperature and pressure from solid ice and salt, or from solid ice and hydrate, there will be an increase of volume.

145. The foregoing investigations enable us to describe the behaviour of a saline solution under various circumstances. This will be made clear by the following examples.

Let the solution be constantly in a state of stable equilibrium: also suppose that initially it is neither saturated with salt nor with ice and that the pressure is so great that there is no danger of evaporation.

Then we may diminish θ without altering p or h until the right line, parallel to $O\theta$, which represents the successive conditions of the solution, meets one of the surfaces

$$\left.\begin{array}{c}\mathcal{F}_w = \Psi_i \\ \mathcal{F}_s = \Psi_s\end{array}\right\}.$$

At that instant a sudden discontinuity occurs. If, for

example, the straight line meets both surfaces on the freezing line of the cryohydrate, it will still be possible to diminish θ without altering p or h, but the solution will be frozen. In any other case when we diminish θ, it will be necessary to vary one or both of the other quantities (p, h). Thus if we wish to keep p constant, it has already been proved that any decrease of θ on the surface $\mathcal{F}_w = \Psi_i$ will make h increase by the freezing out of pure ice until the freezing line of the cryohydrate is reached: on the surface $\mathcal{F}_s = \Psi_s$, it appears from experiment that when θ is diminished, the solution generally deposits the anhydrous salt or a hydrate, so that h decreases. If while θ diminishes, we keep h constant and vary p, we shall never arrive at the freezing line of the cryohydrate, because along that line h has a different constant value.

If we begin by keeping θ and h constant and diminish p, the successive conditions of the solution will be represented by a straight line parallel to Op until we come to one of the three surfaces

$$\left.\begin{array}{l}\mathcal{F}_s = \Psi_s \\ \mathcal{F}_w = \Psi_i \\ \mathcal{F}_w = \Psi_w\end{array}\right\}.$$

It will be sufficient to indicate by an illustration what may happen after we get to the surface $\mathcal{F}_w = \Psi_w$. For simplicity, let us suppose that there is some vapour present with the liquid solution.

Then, since $\dfrac{d_p h}{dh}$ has been shown to be positive on the surface $\mathcal{F}_w = \Psi_w$, we must diminish h at the same time as θ in order to keep p constant. This may be done by causing the volume to decrease so that some of the vapour is condensed.

On the curve of intersection of $\mathcal{F}_w = \Psi_w$ with $\mathcal{F}_w = \Psi_i$, pure ice will first appear; and on the curve of intersection with $\mathcal{F}_s = \Psi_s$, the solution will begin to deposit the salt or a hydrate and may be dissociated by evaporation at constant temperature and pressure.

NOTE A.

It is often supposed that to every action there is simultaneously an equal and opposite reaction, whether these actions be contact-forces or actions at a distance. We have assumed this to be the case with contact-forces, and it is likewise undoubtedly true for the simpler actions at a distance, such as the gravitational and electro-statical and magneto-statical forces between bodies at rest in an unvarying state. For example, consider the mutual gravitational influence of two distant bodies A, B, which are devoid of electric and magnetic properties. If both A and B be at rest, it is quite reasonable to suppose that their mutual influence consists of a set of elementary forces which are *strictly* equal and opposite in pairs. If, however, while B is kept at rest, A be suddenly moved nearer or further away with great velocity, it is clear that the gravitational force on A will at once begin to alter, while that on B will remain unchanged until the effect of the motion of A has had time to cross the space AB—unless, indeed, we can conceive the gravitational force which acts on B capable of foreseeing the intentions of the machine by which the body A is moved.

If we assume, as is usually done, that when the two

bodies have been held at rest for a sufficient time in an unvarying state at a given distance apart and in a given relative position, the energy of the system of the two bodies is independent of the previous history of the system, we shall be led to some important conclusions. For let the system be brought from any such condition (1) to any other such condition, (2) in two different ways, one of which is practically reversible and the other very rapid. Also suppose the machines by which the two bodies are held incapable of doing any but mechanical work. Then since the change of energy is the same in the two different methods and the mechanical work obtained from the system is different, it follows that different thermal processes must have taken place in the two operations, and clearly the difference can only be due to *radiation*.

Again, let A, B be two small distant bodies of any the same uniform temperature θ, which may possess any electric and magnetic properties but cannot gain or lose electric energy, and let them be in such an unvarying state and position that the energy of the system which they compose is independent of the previous history of the system. Let this system be rotated like a rigid body, in a reversible manner, without varying the temperature, about an axis passing through A. Then if Q be the heat absorbed and W the mechanical work done on the system, we have clearly

$$Q + W = 0,$$

and
$$\frac{Q}{\theta} = 0.$$

Hence $W = 0$, and therefore the action at a distance exerted by A on B must act in the line AB.

Similarly, the action at a distance exerted by B on A must act in the same straight line AB.

Again, if the distance AB be increased or diminished by a given amount in two isothermal reversible ways, first, by moving A along the line AB while B is kept at rest, and secondly, by moving B along AB while A is kept at rest, and if we make the reasonable supposition that the changes of energy and entropy are the same in both ways, it follows that the actions at a distance between A and B, which, we have seen, act in the same straight line AB, are equal and opposite.

If the two bodies A, B can gain or lose electric energy, the actions at a distance between them are not necessarily equal and opposite; but since it is always immaterial whether A be moved in one direction or B an equal distance with the same velocity in an opposite parallel direction, it is easily seen that the actions at a distance are equal and in opposite parallel directions.

NOTE B.

We have already noticed a point in which the nomenclature of Thermodynamics differs from that which is unfortunately adopted in books on Rigid Dynamics. We now propose to consider briefly the chief problem discussed in these books.

It has been shown that if a body which possesses angular momentum be left to itself, it cannot move as rigid unless the axis of rotation through G, the centre of mass, be fixed in the body and coincide with the axis of resultant angular momentum through G, which is a line whose directions are fixed in space. Nevertheless many bodies are so hard and unyielding that we may often suppose, without *serious* error, that, for a longer or shorter time, they move as rigid *whatever be their axes of rotation*. This case of motion alone is considered in books on Rigid Dynamics.

We have seen that when any body which possesses angular momentum is left to itself, it will ultimately move as rigid and that the ultimate axis of rotation through the centre of mass will be fixed in the body and coincide with the axis of resultant angular momentum through G, or be a *principal axis* at G. This is true

NOTE B. 383

whether the body be hard or soft; only in the former case, the principal axis will be practically ready-made, while in the latter, it will be formed to a greater or less extent by the motion of the body itself. But since we can always conceive a hard and nearly rigid body having its mass distributed in exactly the same way as any given soft body, it is clear that a principal axis always exists ready-made at G, however soft or mobile the body may be.

Suppose Gz to be a principal axis at G and let it be perpendicular to the plane of the paper. Also let Gx, Gy be any two rectangular axes in the plane of the paper. Then if (x, y, z) be the coordinates of any particle, m, of the body, with respect to these axes, the necessary and sufficient condition that Gz should be a principal axis, is

$$\Sigma mzx = \Sigma mzy = 0.$$

Now if Gx', Gy' be any other pair of rectangular axes in

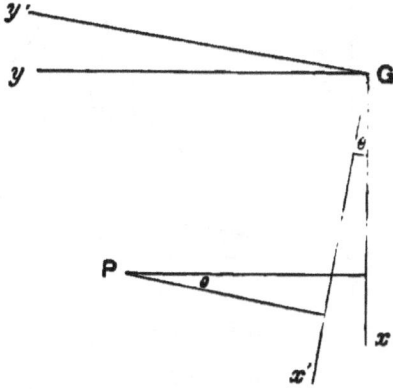

the plane of the paper, making an angle θ with the first pair, we shall have

$$\left. \begin{array}{l} x' = x \cos \theta + y \sin \theta \\ y' = y \cos \theta - x \sin \theta \end{array} \right\},$$

and therefore
$$\Sigma mx'y' = \cos 2\theta \Sigma mxy - \sin\theta \cos\theta \Sigma m(x^2 - y^2).$$
Thus whatever Σmxy and $\Sigma m(x^2 - y^2)$ may be, we can always choose two rectangular axes Gx', Gy', such that $\Sigma mx'y' = 0$. In this case, the three rectangular axes Gx', Gy', Gz will all be principal axes.

Now let Gz' be at any instant the axis of rotation and ω the angular velocity, of a body which is moving as rigid, and let Gz be any other axis through G making an angle θ with Gz'. Take Gx', Gx in the plane $z'Gz$, at right angles to Gz', Gz, respectively, and let Gy' be perpendicular to this plane. Then since the angular momentum about Gz' is $\omega \Sigma m(x'^2 + y'^2)$, and about Gx', $-\omega \Sigma mz'x'$, the angular momentum about Gz will be
$$\omega \cos\theta \Sigma m(x'^2 + y'^2) + \omega \sin\theta \Sigma mz'x'.$$

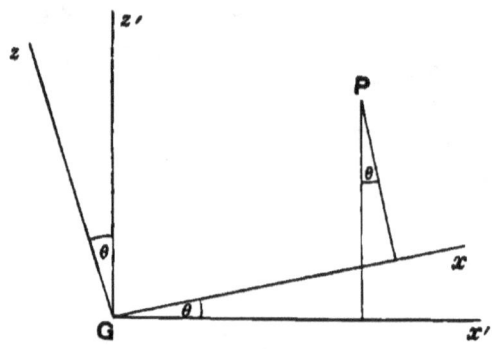

But since
$$z' = x \sin\theta + z \cos\theta \\ x' = x \cos\theta - z \sin\theta$$,
this becomes
$$\omega \cos\theta \Sigma m(x^2 \cos^2\theta + z^2 \sin^2\theta + y'^2)$$
$$+ \omega \sin^2\theta \cos\theta \Sigma m(x^2 - z^2)$$
$$+ \omega \{-2\sin\theta \cos^2\theta + \sin\theta \cos 2\theta\} \Sigma mxz,$$
or
$$\omega \cos\theta \Sigma m(x^2 + y^2) - \omega \sin\theta \Sigma mxz.$$

If Gz be a principal axis at G, $\Sigma mxz = 0$, and the angular momentum about Gz reduces to $\omega \cos \theta \Sigma m (x^2 + y'^2)$. This result will appear to be very simple as soon as we have proved that angular rotations may be resolved and compounded like forces, so that $\omega \cos \theta$ may be called the resolved angular velocity about Gz.

Suppose that angular velocities exist in succession for the same short time τ about two straight lines OS, OT, passing through a fixed point O, and let them be represented in magnitude and direction by the lengths OA, OB. Complete the parallelogram AOB, and from any point P in the diagonal OC drop perpendiculars PM, PN on OS

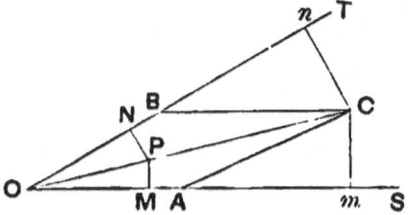

and OT. Then an angular velocity about OS proportional to OA will cause the point P to move in a short time τ perpendicular to the plane SOT a distance proportional to $OA \cdot PM \cdot \tau$. If, after the time τ, the angular rotation about OS cease, and a new angular velocity proportional to OB begin about the new position of OT, the point P will, in the short time τ, be brought back perpendicular to the plane of the paper a distance proportional to $OB \cdot PN \cdot \tau$. Now if Cm, Cn be the perpendiculars from C on OS and OT, we have $PM : PN = Cm : Cn$, and therefore $OA \cdot PM : OB \cdot PN = OA \cdot Cm : OB \cdot Cn = 1 : 1$, since the triangles COA, COB are equal. Hence the point P is brought back to its original position by the second rotation, and therefore, since any displacement of a rigid body with one point, O, fixed, can be effected by a rotation about an axis through O, the two rotations OA, OB, are equivalent to a single rotation about the diagonal OC, or angular velocities are compounded according to the parallelogramic law.

Again, let a body be moving as rigid and take three principal axes $G\xi$, $G\eta$, $G\zeta$, fixed in the body, as rectangular axes of coordinates. Let the moments of inertia of the body about these axes be denoted by A, B, C, which will, of course, be constant quantities however the axes may move about in space. Then if, at any instant, $(\omega_1, \omega_2, \omega_3)$ be the resolved angular velocities about these axes, $(A\omega_1, B\omega_2, C\omega_3)$ will be the resolved angular momenta. Hence if (α, β, γ) be the angles the axis of rotation makes with the axes of coordinates, the resolved angular momentum about the axis of rotation will be

$$A\omega_1 \cos \alpha + B\omega_2 \cos \beta + C\omega_3 \cos \gamma.$$

To get the mechanical kinetic energy of rotation, we multiply this by $\frac{1}{2}\omega$. It is therefore equal to

$$\tfrac{1}{2}(A\omega_1^2 + B\omega_2^2 + C\omega_3^2).$$

If t be the time at which the angular velocities are $(\omega_1, \omega_2, \omega_3)$, then at the consecutive instant $t + dt$, the body will have angular velocities $\omega_1 + \frac{d\omega_1}{dt} dt$, &c., &c., about the new positions of $G\xi$, $G\eta$, $G\zeta$, and angular momenta

$$A\left(\omega_1 + \frac{d\omega_1}{dt} dt\right), \text{ &c., &c.}$$

If we denote the position of the axes at the time t by $G\xi$, $G\eta$, $G\zeta$, and at the time $t + dt$, by $G\xi'$, $G\eta'$, $G\zeta'$, the angles between $G\xi$ and $G\eta'$, $G\zeta'$, will be practically $\frac{\pi}{2} + \omega_3 dt$ and

$\frac{\pi}{2} - \omega_2 dt$. Hence the angular momentum of the body at the time $t + dt$ about $G\xi$ will be

$$A\left(\omega_1 + \frac{d\omega_1}{dt} dt\right) - (B - C)\omega_2\omega_3 dt,$$

so that if, at the time t, L be the moment of the external forces about $G\xi$, we shall have

$$A\frac{d\omega_1}{dt} - (B - C)\omega_2\omega_3 = L.$$

Similarly, if M, N be the moments of the forces about the other axes at the time t,

$$B\frac{d\omega_2}{dt} - (C - A)\omega_3\omega_1 = M,$$

and

$$C\frac{d\omega_3}{dt} - (A - B)\omega_1\omega_2 = N.$$

These three equations are due to Euler, and are known as Euler's equations.

If we take three rectangular axes Ox, Oy, Oz, fixed in space, and denote the sums of the resolved parts parallel to these axes of the external forces which act on the body by P, Q, R, and the coordinates of G by (x, y, z), we also have

$$\left. \begin{aligned} M \frac{d^2x}{dt^2} &= P \\ M \frac{d^2y}{dt^2} &= Q \\ M \frac{d^2z}{dt^2} &= R \end{aligned} \right\},$$

where M is the mass of the body.

When L, M and N are zero, we obtain, by multiplying Euler's equations by ω_1, ω_2, ω_3, respectively, adding, and integrating,

$$\tfrac{1}{2} (A\omega_1^2 + B\omega_2^2 + C\omega_3^2) = \text{constant},$$

or the mechanical kinetic energy of rotation is constant.

When P, Q and R are zero, we find in a similar manner,

$$\tfrac{1}{2} M \left\{ \left(\frac{dx}{dt}\right)^2 + \left(\frac{dy}{dt}\right)^2 + \left(\frac{dz}{dt}\right)^2 \right\} = \text{constant},$$

or the mechanical kinetic energy of translation is constant.

When there are no external forces at all, the total mechanical kinetic energy is constant.

Now we have seen that when a body of which no part of the energy can vary but the mechanical kinetic energy, is left to itself, the mechanical kinetic energy continually decreases, except when the axis of rotation is a principal axis. The results which have been just obtained cannot therefore be strictly true except when the axis of rotation is a principal axis, or two of the three quantities ($\omega_1, \omega_2, \omega_3$) zero. Still they may often be very close approximations for a short time.

To find the mechanical work done by the external forces, we take account only of the mechanical displacements of the points of application. As in books on Rigid Dynamics, we shall restrict ourselves to the case in which the rods &c. by which the external forces are applied, do not slip on the surface of the body—a very simple restriction which is usually expressed in mysterious unphysical language. The mechanical displacement of the point of application of any one of the forces will then be the same as that of some point of the body itself.

Through O draw three fixed rectangular axes Ox', Oy', Oz', parallel to the moving axes $G\xi$, $G\eta$, $G\zeta$ at the time t. Then if (x', y', z') be the coordinates of G at the time t with respect to these axes, and (ξ, η, ζ) the coordinates of a point of the body with respect to the moving axes, the mechanical displacement of this point in the short time dt will have a resolved part parallel to Ox' equal to

$$\left\{\frac{dx'}{dt} + (\omega_2 \zeta - \omega_3 \eta)\right\} dt.$$

Hence since the mechanical work done by a force is equal to the sum of the mechanical works done by its components, if (X', Y', Z') be the resolved parts, parallel to Ox', Oy', Oz', of the force which acts at (ξ, η, ζ), the mechanical work which this force does in the time dt bears to dt the ratio

$$X'\frac{dx'}{dt} + Y'\frac{dy'}{dt} + Z'\frac{dz'}{dt} + \omega_1(Z'\eta - Y'\zeta) + \omega_2(X'\zeta - Z'\xi) + \omega_3(Y'\xi - X'\eta).$$

Thus, since
$$\Sigma(Z'\eta - Y'\zeta) = L$$
$$\Sigma(X'\zeta - Z'\xi) = M,$$
$$\Sigma(Y'\xi - X'\eta) = N$$

the mechanical work done in the time dt by all the forces will be

$$dx'\Sigma X' + dy'\Sigma Y' + dz'\Sigma Z' + (L\omega_1 + M\omega_2 + N\omega_3)\,dt,$$

which is clearly equal to

$$P\,dx + Q\,dy + R\,dz + (L\omega_1 + M\omega_2 + N\omega_3)\,dt.$$

Substituting for P, Q, R, L, M, N, their values as already found, the expression for the elementary mechanical work becomes

$$\mathcal{M}\left(\frac{d^2x}{dt^2}\frac{dx}{dt} + \ldots + \ldots\right)dt + \left(A\frac{d\omega_1}{dt}\omega_1 + \ldots + \ldots\right)dt,$$

or simply dT,

where T is the mechanical kinetic energy.

Denoting the elementary mechanical work by dW, we have

$$dT = dW,$$

and we see that the mechanical work done on the body in any finite time is equal to the corresponding increase of the mechanical kinetic energy. This is the only case of the principle of energy which is considered in books on Rigid Dynamics and it is, in general, only a near approximation.

This result, which has been obtained as a deduction from the laws of motion, is obvious if we assume the general principle of energy. For since the only part of the energy of the body which can vary is the mechanical kinetic energy, the increase of energy in any short interval is dT, and since the only kind of work done on the body is mechanical work, the total work done on it in the short interval is dW. Hence, as before, $dT = dW$.

It will be readily understood that dW need not be a perfect differential; nevertheless the impression generally produced by the ordinary books on Rigid Dynamics is that the principle of energy requires that dW should always be a perfect differential. All that is meant is the following. The given body is supposed to form part of one system with the rods, bars, &c., and external gravitating masses, to which the external forces are due. The new system is supposed to have no external forces acting on it, nor to receive or lose heat, and it is further supposed that the only parts of its energy which can vary are the mechanical kinetic energy of the given body and the mutual potential energy, V, of the given body and the rest of the system. The principle of energy then gives

$$T + V = \text{constant},$$

so that dW is the complete differential of the function $-V$.

As we have already said, the books on Rigid Dynamics unfortunately call V the potential energy of the given body, instead of the mutual potential energy of the given body and the rest of the system.

APPENDIX.

Densities of Gases.

(In the last two columns, the pressure is supposed to be one atmo and the temperature 0° C.)

	Relative densities.	Relative specific volumes.	Mass of a litre in grammes.	Volume of a gramme in litres.
Air	1	1	1·2932	·7733
Oxygen (O)	1·10563	·90446	1·4298	·6994
Hydrogen (H)	·06926	14·4383	·08957	11·16445
Nitrogen (N)	·97135	1·02945	1·25615	·7961
Carbonic Oxide (CO)	·9545	1·0476	1·2344	·8101
Carbonic Acid (CO_2)	1·52907	·6540	1·9774	·5057
Chlorine (Cl)	2·4222	·4128	3·1328	·3192
Cyanogen (NC_2)	1·8019	·5550	2·3302	·4291
Marsh Gas (CH_4)	·562	1·779	·727	1·375
Olefiant Gas (C_2H_4)	·982	1·018	1·270	·787
Ammonia (NH_3)	·5952	1·6801	·7697	1·2992

Specific Heats of Gases.

	At constant pressure		At constant volume		Ratio of the specific heats, or k.
	In calories. (By experiment.)	Compared with an equal vol. of air.	In calories.	Compared with an equal vol. of air.	
Air	·2375	1	·1684	1	1·410
Oxygen	·21751	1·012	·15501	1·018	1·403
Hydrogen	3·40900	·994	2·4114	·992	1·414
Nitrogen	·24380	·997	·17266	·996	1·412
Carbonic Oxide	·2450	·985	·1728	·978	1·418
Carbonic Acid	·2169	1·396	·1717	1·559	1·263
Chlorine	·12099	1·234	·0925	1·330	1·308
Marsh Gas	·5929	1·403	·4700	1·568	1·260
Olefiant Gas	·4040	1·670	·3337	1·946	1·211
Ammonia	·5084	1·274	·3923	1·386	1·296

Water at a pressure of one atmo.

(Everett's 'Units and Physical Constants,' also 'Encyclopedia Britannica'.)

Temperature in degrees centigrade.	Mass of one cubic centimetre in grammes, or of one cubic decimetre in kilogrammes.	Temperature.	Volume of one gramme in cubic centimetres, or of one kilogramme in cubic decimetres.
0°	·999,884	0°	1·000,116
4°	1·000,013	4°	·999,987
5°	1·000,003	5°	·999,997
10°	·999,760	10°	1·000,240
15°	·999,173	15°	1·000,828
20	·998,272	20°	1·001,731
25°	·997,108	25°	1·002,900
30°	·995,778	30°	1·004,240
35°	·994,69	35°	1·005,34
40°	·992,36	40°	1·007,70
45°	·990,38	45°	1·009,71
50°	·988,21	50°	1·011,93
55°	·985,83	55°	1·014,37
60°	·983,39	60°	1·016,89
65°	·980,75	65°	1·019,63
70°	·977,95	70°	1·022,55
75°	·974,99	75°	1·025,65
80°	·971,95	80°	1·028,86
85°	·968,80	85°	1·032,20
90°	·965,57	90°	1·035,66
95°	·962,09	95°	1·039,40
100°	·958,66	100°	1·043,12

According to Regnault, the specific heat of water (in calories) is as follows:

at 0° C.	1·0000
at 10° ,,	1·0005
at 20° ,,	1·0012
at 30° ,,	1·0020

APPENDIX. 395

Ice at the pressure of one atmo and at 0° C.

Mass of one cubic centimetre = ·920 gramme.
Volume of one gramme = 1·087 cubic centimetres.
Coefficient of cubical dilatation by heat at constant pressure = ·000153.
Specific heat (in calories) at constant pressure = ·48.

(Person.)

Mercury at a pressure of one atmo (or less).

It has been found (Everett's 'Units') that the density of mercury at 0° C. is 13·5956 times that of water at 4° C. It is therefore 13·595 776 74 grammes per cubic centimetre. Hence the pressure produced (at Paris) by a column of mercury at 0° C. and one millimetre high, whose top is acted on by no force but the insignificant pressure of its own saturated vapour, is 1333·5662 dynes per square centimetre. Also an atmo, or the pressure produced (at Paris) by a column of mercury at 0° C. and 760 millimetres high, is 1,013,510·3356 dynes, or 1033·279 grammes, per square centimetre.

Melting Points and Latent Heats of Fusion (in calories) of Solids at a pressure of one atmo.

(From Watt's 'Dictionary of Chemistry.')

	Melting-points (C.).	Latent Heats.
Mercury	−39	2·82
Phosphorus	44°·2	5·0
Sulphur	115°	9·4
Iodine	107°	11·7
Lead	332°	5·4
Tin	235°	14·25
Silver	1000°	21·1
Zinc	433°	28·1
Bismuth	270°	12·6
Nitrate of Potassium	339°	47·4
Nitrate of Sodium	310°·5	63·0

Boiling Points and Heats of Vaporisation at a pressure of one atmo.

(Everett's 'Units'.)

	Boiling-points (C.).	Latent Heat of Vaporisation.	Observer.
Alcohol	77°·9	202·4	Andrews
Bisulphide of Carbon	46°·2	86·7	,,
Bromine	58°	45·6	,,
Ether	34°·9	90·4	,,
Mercury	350° (?)	62	Person
Sulphur	316° (?)	362	,,
also			
Sulphurous Acid	−10°·08		
Concentrated Sulphuric Acid	325°		

Temperatures and Pressures of Critical Points.

(Cagniard de la Tour.)

	Critical Temperature.	Pressure in atmos.
Bisulphide of Carbon	262·5° C.	66·5
Ether	187·5° C.	37·5
Alcohol	258·7° C.	119
Water	411·7° C.	?

In the case of water, Maxwell estimates the Critical Temperature to be about 434° C., the Critical Pressure about 378 atmos, and the Critical Volume about 2·52 cubic centimetres per gramme.

Pressure of the saturated vapour of Water
(calculated from Regnault).

[When the pressure is measured by a column of mercury, the mercury is supposed to be contained in a wide (non-

APPENDIX. 397

capillary) tube at Paris, at 0° C., and subjected to no surface pressure beyond the negligible pressure of its own vapour.]

Pressures.

Temperatures (C.).	In millimetres of mercury.	In grammes (at Paris) per square centimetre.	In dynes per square centimetre.	In atmos.	In pounds per sq. inch (at London).
−30°	·4	·54	529·7		·008
−25°	·6	·82	804·3		·012
−20°	·9	1·22	1,196·7	·001	·017
−15°	1·4	1·90	1,863·6	·002	·027
−10°	2·1	2·85	2,795·5	·003	·041
−5°	3·1	4·21	4,129·4	·004	·059
0°	4·600	6·254	6,134·40	·006	·089
5°	6·534	8·883	8,713·52	·0086	·126
10°	9·165	12·460	12,222·1	·0120	·177
15°	12·699	17·265	16,934·9	·0168	·246
20°	17·391	23·644	23,192·0	·0229	·336
25°	23·550	32·018	31,405·5	·0309	·455
30°	31·548	42·892	42,071·3	·0415	·610
35°	41·827	56·867	55,779·1	·0550	·809
40°	54·906	74·649	73,220·8	·0722	1·062
45°	71·390	97·060	95,203·3	·0939	1·380
50°	91·980	125·054	122,661	·1210	1·779
55°	117·475	159·716	156,661	·1545	2·27
60°	148·786	202·262	198,416	·1957	2·88
65°	186·938	254·247	249,294	·246	3·61
70°	233·082	316·893	310,830	·306	4·51
75°	288·500	392·238	384,734	·379	5·58
80°	354·616	482·128	472,904	·467	6·86
85°	433·002	588·700	577,437	·570	8·37
90°	525·392	714·311	700,645	·704	10·16
95°	633·692	861·553	845,070	·833	12·25
100°	760	1033·279	1,013,510	1	14·697
105°	906·41	1232·33	1,208,760	1·19	17·53
110°	1075·37	1462·05	1,434,080	1·41	20·80
115°	1269·41	1725·86	1,692,840	1·67	24·55
120°	1491·28	2027·51	1,988,720	1·96	28·84
125°	1743·88	2370·94	2,325,580	2·29	33·72
130°	2030·28	2760·32	2,707,510	2·67	39·26
135°	2353·73	3200·08	3,138,850	3·09	45·5
140°	2717·63	3694·83	3,624,140	3·57	52·6
145°	3125·55	4249·43	4,168,130	4·11	60·4
150°	3581·23	4868·97	4,775,810	4·71	69·3
155°	4088·56	5558·71	5,452,370	5·38	79·1

Temperatures (C.).	In millimetres of mercury.	In grammes (at Paris) per square centimetre.	In dynes per square centimetre.	In atmos.	In pounds per sq. inch (at London).
160°	4651·62	6324·24	6,203,240	6·12	90·0
165°	5274·54	7171·15	7,033,950	6·94	102·0
170°	5961·66	8105·34	7,950,270	7·84	115·3
175°	6717·43	9132·87	8,958,140	8·84	129·9
180°	7546·39	10259·9	10,063,600	9·93	145·9
185°	8453·23	11492·8	11,272,900	11·12	163·5
190°	9442·70	12837·1	12,592,500	12·42	182·6
195°	10519·63	14302·2	14,028,600	13·84	203·4
200°	11688·96	15892·0	15,588,000	15·38	226·0
205°	12955·7	17614·3	17,277,300	17·0	250·5
210°	14324·8	19475·7	19,103,100	18·8	277·0
215°	15801·3	21483·1	21,072,100	20·8	305·6
220°	17390·4	23643·6	23,191,200	22·9	336·3
225°	19097	25963·8	22,800,000	25·1	369·3
230°	20926·4	28451·1	27,906,700	27·5	404·7

Pressure of saturated vapour of Mercury (in millimetres of mercury).

(From the 'Encyclop. Brit.')

Temperatures (C.).	Pressures.	Temperatures (C.).	Pressures.	Temperatures (C.).	Pressures.
0°	·02	180°	11	360°	797·7
10°	·03	190°	14·8	370°	954·6
20°	·04	200°	19·9	380°	1139·6
30°	·05	210°	26·3	390°	1346·7
40°	·08	220°	34·7	400°	1588
50°	·11	230°	45·3	410°	1863·7
60°	·16	240°	58·8	420°	2177·5
70°	·24	250°	75·7	430°	2533
80°	·35	260°	96·7	440°	2934
90°	·51	270°	123	450°	3384·4
100°	·75	280°	155·2	460°	3888·1
110°	1·07	290°	194·5	470°	4449·4
120°	1·53	300°	242·2	480°	5072·4
130°	2·18	310°	299·7	490°	5761·3
140°	3·06	320°	368·7	500°	6520·3
150°	4·27	330°	450·9	510°	7253·4
160°	5·90	340°	548·3	520°	8265
170°	8·09	350°	663·2		

APPENDIX.

Pressure of saturated vapour of Sulphur (in millimetres of mercury).
('Encyclop. Brit.')

Temps. (C.)	Pressures.	Temps. (C.)	Pressures.	Temps. (C.)	Pressures.
390°	272·3	460°	912·7	520°	2133·3
400°	329	470°	1063·2	530°	2422
410°	395·2	480°	1232·7	540°	2739·2
420°	472·1	490°	1422·9	550°	3086·5
430°	561	500°	1635·3	560°	3465·3
440°	663·1	510°	1871·6	570°	3877·1
450°	779·9				

Pressures of saturated vapours (in millimetres of mercury).
('Encyclop. Brit.')

Temps. (C.)	Ammonia. (NH_3).	Sulphuretted Hydrogen. (H_2S).	Carbonic Acid. (CO_2).	Nitrous Oxide. (N_2O).
−30°	866·1			
−25°	1104·3	3749·3	13007	15694·9
−20°	1392·1	4438·5	15142·4	17586·6
−15°	1736·5	5196·5	17582·5	19684·3
−10°	2144·6	6084·6	20340·2	22008
−5°	2624·2	7066	23441·3	24579·2
0°	3183·3	8206·3	26906·6	27421
5°	3830·3	9490·8	30753·8	30558·6
10°	4574	10896·3	34998·6	34019·1
15°	5423·4	12447·9	39646·9	37831·7
20°	6387·8	14151·5	44716·6	42027·9
25°	7477	16012·4	50207·3	46641·4
30°	8701	18035·3	56119	51708·5
35°	10070·2	20224·3	62447·3	57268·1
40°	11595·3	22582·5	69184·4	63359·8
45°	13287·3	24954·3	76314·6	
50°	15158·3	27814·8		
55°	17219·8	30690·7		
60°	19482·1	33740·2		
65°	21965·1	36961·5		
70°	24675·5	40353·2		
75°	27630			
80°	30843·1			
85°	34330·9			
90°	38109·2			
95°	42195·7			
100°	46608·2			

APPENDIX.

Pressures of saturated vapours (in millimetres of mercury).
('Encyclop. Brit.')

Temps. (C).	Essence of Turpentine. ($C_{10}H_6$).	Chloroform. ($CHCl_3$).	Carbon Bisulphide. (CS_2).	Sulphurous Acid. (SO_2).
−30°				287·5
−25°				373·8
−20°			47·3	479·5
−15°			61·6	607·9
−10°			79·4	762·5
−5°			101·3	946·9
0°	2·1		127·9	1165·1
5°			160	1421·1
10°	2·9		198·5	1719·5
15°			244·1	2064·9
20°	4·4	160·5	298	2462
25°		200·2	361·1	2916
30°	6·9	247·5	434·6	3431·8
35°		303·5	519·7	4014·8
40°	10·8	369·3	617·5	4670·2
45°		446	729·5	5403·5
50°	17	535	857·1	6220
55°		637·7	1001·6	7125
60°	26·5	755·4	1164·5	8123·8
65°		889·7	1347·5	9221·4
70°	40·6	1042·1	1552·1	
75°		1214·2	1779·9	
80°	61·3	1407·6	2032·5	
85°		1624·1	2311·7	
90°	90·6	1865·2	2619·1	
95°		2132·8	2966·3	
100°	131·1	2428·5	3325·1	
105°		2754	3727·2	
110°	186	3111	4164·1	
115°		3501	4637·4	
120°	257	3925·7	5148·8	
125°		4386·6	5699·7	
130°	349	4885·1	6291·6	
135°		5422·5	6925·9	
140°	464	6000·2	7604	
145°		6619·2	8326·9	
150°	605	7280·6	9095·9	
155°	686	7985·3		
160°	775	8734·2		
165°		9527·8		

APPENDIX.

Solution of Gases in Water.
(From Roscoe and Schorlemmer's 'Chemistry'.)

The number of volumes of gas (reduced to 0° C. and one atmo) absorbed by one volume of water, is called the 'coefficient of solubility' of the gas.

Oxygen (O).

The coefficient of solubility C, at the temperature θ°C., is given by
$$C = \cdot 04115 - \cdot 0010899\,\theta + \cdot 000022563\,\theta^2.$$

Hydrogen (H).

Unlike most other gases, Hydrogen is equally soluble for all temperatures between 0°C. and 20°C., the coefficient of solubility being $\cdot 0193$.

Nitrogen (N).

The coefficient of solubility is
$$C = \cdot 020346 - \cdot 00053887\,\theta + \cdot 00001156\,\theta^2.$$

Sulphuretted Hydrogen (H_2S).

Between 0°C. and 40°C., the coefficient of solubility is
$$C = 4\cdot 3706 - \cdot 083687\,\theta + \cdot 0005213\,\theta^2.$$

Carbonic Acid (CO_2).

Specific gravity of liquid Carbonic Acid is $\cdot 9951$ at -10°C., $\cdot 9470$ at 0°C., and $\cdot 8266$ at 20°C. It is thus very expansible by heat.

The boiling point under a pressure of one atmo is $-78\cdot 2$°C. For the coefficient of solubility we have the formula

$$C = 1{\cdot}7967 - {\cdot}07761\theta + {\cdot}0016424\theta^2,$$

or

$$\left.\begin{array}{l}\text{at } 0°, \; C = 1{\cdot}7967 \\ \phantom{\text{at }} 5°, \; C = 1{\cdot}4497 \\ \phantom{\text{at }} 10°, \; C = 1{\cdot}1848 \\ \phantom{\text{at }} 15°, \; C = 1{\cdot}0020 \\ \phantom{\text{at }} 20°, \; C = {\cdot}9014\end{array}\right\}..$$

When the pressure is much smaller than that of the atmosphere, the quantity of gas absorbed at a given temperature is proportional to the pressure.

Chlorine (Cl).

The coefficient of solubility between 10° C. and 40° C. is given by the formula

$$C = 3{\cdot}0361 - {\cdot}046196\theta + {\cdot}0001107\,\theta^2.$$

Hence

$$\left.\begin{array}{l}\text{at } 10°, \; C = 2{\cdot}5852 \\ \phantom{\text{at }} 20°, \; C = 2{\cdot}1565 \\ \phantom{\text{at }} 30°, \; C = 1{\cdot}7499 \\ \phantom{\text{at }} 40°, \; C = 1{\cdot}3654\end{array}\right\}.$$

Ammonia (NH_3).

The coefficient of dilatation by heat of liquid ammonia at 0° C. is ·00204: the specific gravity at the same temperature compared with water is ·6234.

It has been found by Roscoe and Dittmar that the quantity of ammonia absorbed at a given temperature is not proportional to the pressure; but the deviations diminish as the temperature increases.

The solubility of ammonia at different temperatures under a pressure of one atmo is shown in the following table, due to Roscoe and Dittmar.

Temps. (C.)	Grammes of gas absorbed by one gramme of water.	Temps. (C.)	Grammes of gas absorbed by one gramme of water.
0°	·875	30°	·403
2°	·833	32°	·383
4°	·792	34°	·362
6°	·751	36°	·343
8°	·713	38°	·324
10°	·679	40°	·307
12°	·645	42°	·290
14°	·612	44°	·275
16°	·582	46°	·259
18°	·554	48°	·244
20°	·526	50°	·229
22°	·499	52°	·214
24°	·474	54°	·200
26°	·449	56°	·186
28°	·426		

Hydrochloric Acid (HCl).

Hydrochloric Acid is very soluble in water, and the solution is frequently called 'muriatic acid.' The quantity of gas dissolved at a given temperature is not proportional to the pressure.

The mass of hydrochloric acid absorbed under the pressure of one atmo by one gramme of water at different temperatures is given by the following table.

Temps. (C.)	Grammes of HCl.	Temps. (C.)	Grammes of HCl.
0°	·825	32°	·665
4°	·804	36°	·649
8°	·783	40°	·633
12°	·762	44°	·618
16°	·742	48°	·603
20°	·721	52°	·589
24°	·700	56°	·575
28°	·682	60°	·561

Dilution of Sulphuric Acid (H_2SO_4) with water.

When Sulphuric Acid is mixed with water, there is a contraction of volume and a considerable evolution of heat. The quantity of heat that must be carried off to keep the temperature constant has been determined by Thomsen, from whose results the following table is derived. We suppose that there is originally one gramme of (HCl) and that a solution is formed by adding x molecules of (H_2O) to each molecule of (HCl), that is, by adding $\frac{18x}{98}$ grammes of (H_2O).

Values of x.	Heat evolved (in calories).	Values of x.	Heat evolved.
1	64	49	170
2	96	99	172
3	113	199	174
5	133	499	175
9	152	799	180
19	166	1599	182

Solution of Salts in water.

The accompanying table exhibits some fundamental properties of salts:

(1) The amount of heat that must be imparted to keep the temperature constant when one gramme of salt is thrown into a large quantity of water. (Favre and Silberman in 'Watt's Dictionary of Chemistry.')

(2) The Specific Heats of salts. (Regnault in 'Watt's Dictionary of Chemistry.')

(3) The approximate Specific Gravities from 'Clarke's Constants of Nature.'

APPENDIX.

Name of Salt.		Heat absorbed.	Specific heat.	Specific gravity.
Sulphate of	Potassium	35·3		2·6
,, ,,	Sodium	49·1		2·65
,, ,,	Calcium	24·7	·19656	
,, ,,	Ammonium	11·1		
,, ,,	Zinc	14·8		
,, ,,	Lead		·08723	6·25
Chloride of	Sodium	8·9	·21401	2·15
,, ,,	Calcium ($CaCl_2$)	15·5	·16420	2·25
,, ,,	Potassium	51·9	·17295	1·98
,, ,,	Magnesium ($HgCl_2$)		·19460	
,, ,,	Zinc		·13618	
,, ,,	Lead		·06641	
Nitrate of	Sodium	45·5	·27821	2·25
,, ,,	Calcium	27·1		
,, ,,	Potassium		·23875	2
Carbonate of	Sodium	52·7	·23115	2·45
,, ,,	Potassium		·19010	2·21

Salts may be divided into two classes: (1) those whose solution deposits the anhydrous salt; (2) those which deposit a hydrate.

To the first class belong:—

> Sulphate of Ammonium
> Chlorate of Potassium
> Chloride of Ammonium
> Nitrate of Potassium
> ,, ,, Lead
> ,, ,, Silver
> Iodide of Potassium

To the second class belong:—

> Sulphate of Ammonium
> ,, ,, Magnesium
> ,, ,, Zinc
> Chloride of Sodium
> ,, ,, Barium

A saturated saline solution always boils at a temperature above 100°C. But the excess of the boiling point above 100°C. is not proportional to the amount of salt dissolved. Thus

	Boiling-point.	100 parts of water dissolve.
Chloride of Sodium	109° C.	38·7
Nitrate of Potassium	114° C.	327·4
Carbonate of Potassium	135° C.	205·0
Iodide of Potassium	118° C.	223·0

The vapour-pressure of an aqueous solution of a salt is always less than the pressure of saturated steam at the same temperature, the difference at any given temperature being roughly proportional to the percentage of salt in the solution. Thus at 51·8°C., the vapour-pressures of aqueous solutions of chloride of sodium, expressed in millimetres of mercury, were found by Wüllner to be:—

(H_2O)	(NaCl)	millims.
100	0	100
90	10	94
80	20	88

Wüllner has also found that the diminution of the vapour-pressure due to the presence of a given percentage of any salt increases with the temperature.

Table shewing the temperatures at which a solid first appears on cooling solutions of chloride of sodium of different strengths. (Dr Guthrie in *Phil. Mag.*, May, 1876.)

NaCl per cent. by weight.	H_2O per cent. by weight.	Temperature (C.) at which a solid formed.	Nature of Solid.
1	99	− ·3°	Ice
2	98	− ·9°	,,
3	97	− 1·5°	,,
4	96	− 2·2°	,,
7	93	− 4·2°	,,

Table (continued).

NaCl per cent. by weight.	H$_2$O per cent. by weight.	Temperature (C.) at which a solid formed.	Nature of Solid.
10	90	$-6\cdot6°$	Ice
13	87	$-9\cdot1°$	,,
15	85	$-11°$	,,
16	84	$-11\cdot9°$	,,
19	81	$-15\cdot5°$	,,
20	80	$-17°$	,,
22	78	$-20°$	,,
23·6	76·4	$-22°$	Cryohydrate
25	75	$-12°$	Bihydrate (NaCl+2H$_2$O)
26·27	73·73	$0°$	,,
26·5	73·5	$+25°$	,,
26·8	73·2	$+40°$	,,

Some of the principal results of Dr Guthrie's experiments on the freezing of saline solutions are given below; showing (1) the lowest temperature which he obtained by means of the cryogen, (2) the freezing point of the cryohydrate (under a pressure of one atmo), (3) the 'water-worth' of the cryohydrate—that is, the number of molecules of water to each molecule of the salt. (*Phil. Mag.*, April, 1875.)

Salt.	Temp. of cryogen (C.)	Temp. of solidification of cryohydrate (C.)	Water-worth of cryohydrate.
Bromide of Sodium (NaBr)	$-28°$	$-24°$	8·1
Iodide of Ammonium (NH$_4$I)	$-27°$	$-27\cdot5°$	6·4
Iodide of Sodium (NaI)	$-26\cdot5°$		5·8
Iodide of Potassium (KI)	$-22°$	$-22°$	8·5
Chloride of Sodium (NaCl)	$-22°$	$-22°$	10·5
Chloride of Strontium (SrCl$_2$+6H$_2$O)	$-18°$	$-17°$	22·9
Sulphate of Ammonium (NH$_4$SO$_4$)	$-17\cdot5°$	$-17°$	10·2
Bromide of Ammonium (NH$_4$Br)	$-17°$	$-17°$	11·1
Nitrate of Ammonium (NH$_4$NO$_3$)	$-17°$	$-17\cdot2°$	5·72
Nitrate of Sodium (NaNO$_3$)	$-16\cdot5°$	$-17\cdot5°$	8·13
Chloride of Ammonium (NH$_4$Cl)	$-16°$	$-15°$	12·4
Bromide of Potassium (KBr)	$-13°$	$-13°$	13·94

Table (*continued*).

Salt.	Temp. of cryogen (C.)	Temp. of solidification of cryohydrate (C.)	Water-worth of cryohydrate.
Chloride of Potassium (KCl)	$-10.5°$	$-11.4°$	16·61
Chromate of Potassium (K_2CrO_4)	$-10.2°$	$-12°$	18·8
Chloride of Barium ($BaCl_2 + 2H_2O$)	$-7.2°$	$-8°$	37·8
Nitrate of Lead {$Pb2(NO_3)$}	$-2.5°$	$-2.5°$	
Nitrate of Strontium (Sr_2NO_3)	$-6°$	$-6°$	33·5
Sulphate of Magnesium ($MgSO_4 + 7H_2O$)	$-5.3°$	$-5°$	23·8
Sulphate of Zinc ($ZnSO_4 + 7H_2O$)	$-5°$	$-7°$	20
Nitrate of Potassium (KNO_3)	$-3°$	$-2.6°$	44·6
(Na_2CO_3)	$-2.2°$	$-2°$	92·75
Sulphate of Copper ($CuSO_4 + 5H_2O$)	$-2°$	$-2°$	43·7
Sulphate of Iron ($FeSO_4 + 7H_2O$)	$-1.7°$	$-2.2°$	41·41
Sulphate of Potassium (K_2SO_4)	$-1.5°$	$-1.2°$	114·2
Bichromate of Potassium (K_2CrO_7)	$-1°$	$-1°$	292
Nitrate of Barium {$Ba2(NO_3)$}	$-·9°$	$-·8°$	259
Crystallized Sulphate of Soda ($Na_2SO_4 + 10H_2O$)	$-·7°$	$-·7°$	165·6
Chlorate of Potassium ($KClO_3$)	$-·7°$	$-·5°$	222
Ammonium Alum {$Al_2NH_4 2(SO_4) + 12H_2O$}	$-·4°$	$-·2°$	261·4
Perchloride of Mercury ($HgCl_2$)	$-·2°$	$-·2°$	450
Nitrate of Silver ($AgNO_3$)	$-6.5°$	$-6.5°$	10·09
Anhydrous Sulphate of Soda (Na_2SO_4)	$-·7°$		
Anhydrous Sulphate of Copper ($CuSO_4$)	$-1.7°$		
Nitrate of Calcium {$Ca2(NO_3)$}	$-16°$		
Chloride of Calcium ($CaCl_2 + 3H_2O$)		$-37°$	11·8

www.ingramcontent.com/pod-product-compliance
Lightning Source LLC
Chambersburg PA
CBHW050844300426
44111CB00010B/1115